Mechanical Engineering Series

For further volumes:
http://www.springer.com/series/1161

The Mechanical Engineering Series features graduate texts and research monographs to address the need for information in contemporary mechanical engineering, including areas of concentration of applied mechanics, biomechanics, computational mechanics, dynamical systems and control, energetics, mechanics ofmaterials, processing, production systems, thermal science, and tribology.

Chakravarti V. Madhusudana

Thermal Contact Conductance

Second Edition

Chakravarti V. Madhusudana
Sydney, NSW
Australia

ISSN 0941-5122 ISSN 2192-063X (electronic)
ISBN 978-3-319-01275-9 ISBN 978-3-319-01276-6 (eBook)
DOI 10.1007/978-3-319-01276-6
Springer Cham Heidelberg New York Dordrecht London

Library of Congress Control Number: 2013943984

Printed on acid-free paper

Springer is part of Springer Science+Business Media (www.springer.com)

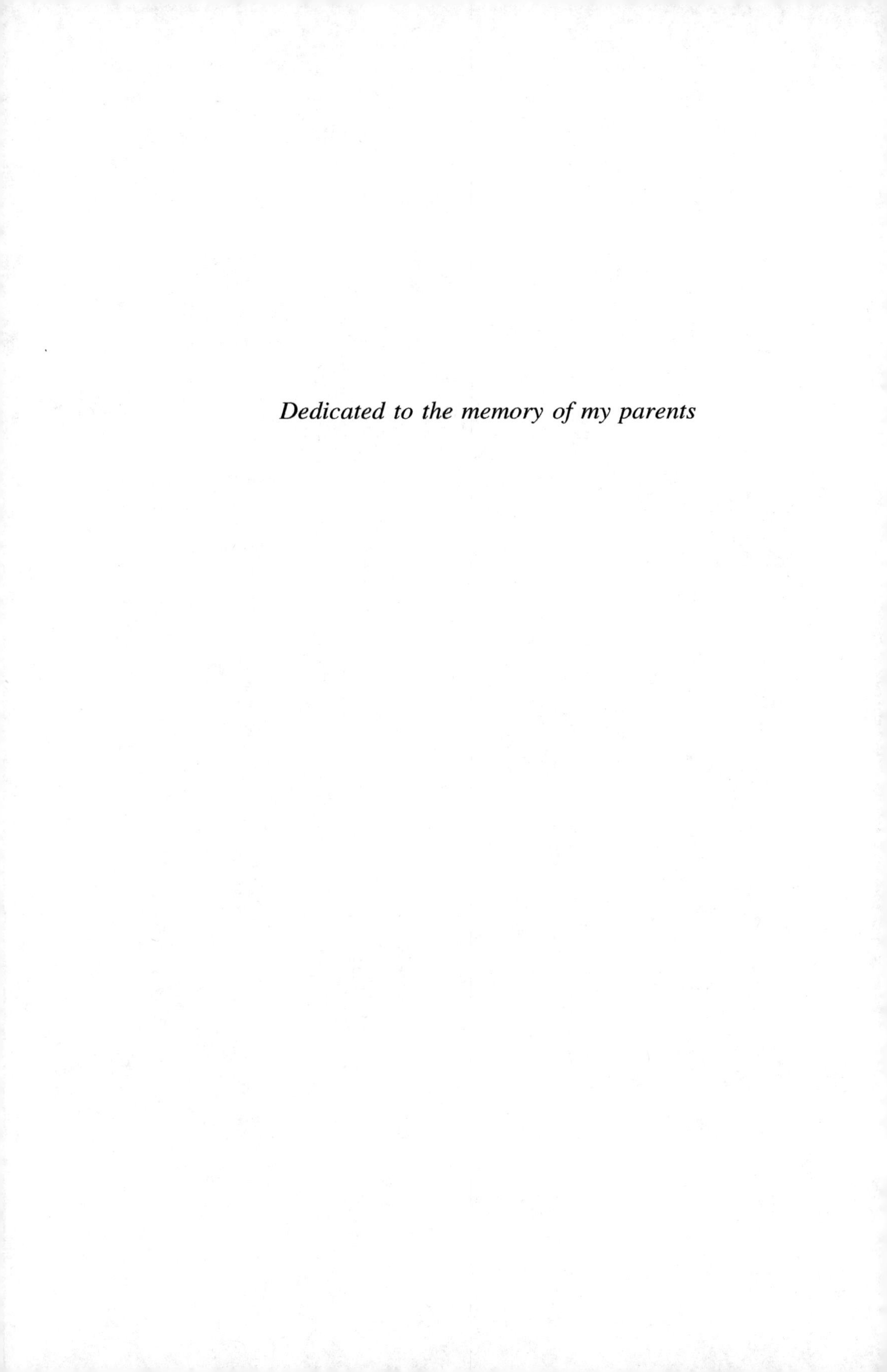

Dedicated to the memory of my parents

Preface to the First Edition

Over the past fifty years, a significant amount of fundamental and applied research has been carried out in the field of contact heat transfer. In this book, I have attempted to synthesise the information generated and present it in a form that is, I hope, interesting to the practising engineer, scientist or student. For this reason, many applications are enumerated in the Introduction and emphasised in the later chapters. It is also hoped that the material presented is readily understandable. To this end, I have explained in some detail the steps involved in developing the basic concepts in Chapters 2 and 3. Although, written mainly for the generalist engineer or physicist, I believe there is sufficient detail, especially in Chapters 4, 6 and 7 to be of interest to the specialist or the advanced student in the field. Chapter 5 should be of particular interest to those contemplating experimental determination of thermal contact conductance and associated problems.

Although more than 500 references were consulted during the preparation of the manuscript, only a few representative ones in each category have been cited, so that the actual number of references listed is approximately half the number consulted. For the sake of systematic and chronological development of each particular theme, however, early pioneering work in that area has necessarily been cited. In general, reference has not been made to internal reports, theses or other work that is not readily available in open literature. Because of language limitations, references in English or English translations of references in other languages, have been used.

Some graphs and diagrams from original sources have been modified and redrawn to suit the format of this work. These have been identified by the words "after" or "based on" followed by the reference to the source.

Since the subject of contact heat transfer has a broad range, it is likely that some topics have been omitted in a work of this size. I hope, however, that the majority of topics, both basic and applied, have received some airing.

It is a pleasure to acknowledge my debt of gratitude to Dr. Arthur Williams who introduced me to contact heat transfer at Monash University. I sincerely appreciate the help and cooperation that I received from Professor Skip Fletcher and Professor Bud Peterson during my sabbaticals at Texas A&M University. Particular mention must be made of the vast source of references that I had access to during those periods. I am grateful to Professor Brian Milton and Professor Mark Wainwright of the University of New South Wales for "reassigning my duties" in lieu of a sabbatical during the first half of 1993. This helped me in gathering, collating and updating the material for this work. Finally, my thanks go to my wife Nagu for her support and understanding.

C. V. Madhusudana

Preface to the Second Edition

More than 17 years have passed since the first edition of *Thermal Contact Conductance* appeared in print. I have taken this opportunity to revise the book in order to reflect the developments in contact heat transfer that have taken place during the intervening years. This revision has also involved reading, reviewing and abstracting from over 150 recent technical papers, reports and theses.

A main feature of this edition is that several new and relevant topics have been added. These include:

Thermal boundary resistance, gap liquid conductance, transient experimental techniques, periodic contacts, heat transfer in sliding friction, carbon nano tubes and other recently developed thermal interstitial materials, finned tube heat exchangers, manufacturing processes, contact heat transfer at low temperatures and non-metallic materials.

About 75 % (96 out of 129) of the diagrams have been specially drawn for this edition.

While the revision has been extensive, it is hoped that the spirit and the format of this edition remain the same as in the first.

My sincere thanks go to Prof. Tomasz Wisniewski for helpful correspondence regarding periodic contacts and permission to use Fig. 6.25.

It is a pleasure to acknowledge the help and support of Springer editorial staff, especially Alex Greene, Ania Levinson and Jessica Lauffer.

As in the previous edition, I wish to thank my wife Nagu for her continuous encouragement and understanding during the preparation of the manuscript.

Chakravarti V. Madhusudana

Acknowledgments

The author thanks the following publishers for granting permission to use copyrighted material:

- American Institute of Physics for

  1. Figure 3.15 (Zhou et al. 2011, Applied Physics Letters, 99:063110)
  2. Figure 7.10 (Son et al. 2008, Journal of Applied Physics, 103:024911)
  3. Figure 8.16 (Bendada et al. 2003, Rev Scientific Instruments, 74:5282–5284)
  4. Figure 9.1 (Mykhyalik et al. 2012, Rev Scientific Instruments, 83:034902)
  5. Figure 9.5 and 9.6 (Wang et al. 2006, Rev Scientific Instruments, 77:024901)

- The American Society of Mechanical Engineers, for Figs. 9.12 and 9.13 (Li et al. 2000, Trans ASME, J Heat Transfer, 122:46–49)

- Elsevier Inc for

  1. Figures 6.10 and 6.11 (Yeh et al. 2001, Experimental Thermal and Fluid Science, 25:349–357)
  2. Figure 8.3 (Jeong et al. 2006, Int J Heat and Mass Transfer, 49:1547–1555)

- Pergamon for

  1. Figure 4.4 (Wahid and Madhusudana, 2000, Int J Heat Mass Transfer, 43:4483–4487)
  2. Figures 6.26 and 6.27 (Wang and Degiovanni 2002, Int J Heat and Mass Transfer 45:2177–2190)
  3. Figure 7.11 (Cola et al. 2009, Int J Heat and Mass Transfer 52 3490–3503)
  4. Figure 7.12 (Shaikh et al. 2007, Carbon, 45:695–703)
  5. Figure 8.25 (Hadley, 1986, Int J Heat Mass Transfer, 29:909–920)
  6. Figures 9.2 and 9.3 (Xu and Xu, 2005, Cryogenics, 45:694–704)

- Springer for the use of data in Table 7.6 (Abadi and Chung, 2011, Journal of Electronic Materials, 40:1490–1500)

Thanks are also due to Copyright Clearance Center for facilitating many of the permissions listed above.

Contents

1 Introduction 1

1.1 Mechanism of Contact Heat Transfer 1

1.2 Significance of Contact Heat Transfer 3

1.3 Scope of Present Work 6

References 7

2 Thermal Constriction Resistance 9

2.1 Circular Disc in Half Space 9

2.2 Resistance of a Constriction Bounded by a Semi-infinite Cylinder 14

2.2.1 Contact Area at Uniform Temperature 15

2.2.2 Contact Area Subjected to Uniform Heat Flux 18

2.3 Eccentric Constrictions 19

2.4 Constriction in a Fluid Environment 19

2.5 Constrictions of Other Types 21

References 22

3 Solid Spot Thermal Conductance of a Joint 25

3.1 Multiple Spot Contact Conductance 25

3.2 Estimation of the Number and Average Size of the Contact Spots 27

3.2.1 Gaussian Distribution of Heights and Slopes 30

3.3 Deformation Analysis 31

3.3.1 The Plasticity Index 31

3.3.2 Ratio of Real to Apparent Area of Contact 32

3.4 Theoretical Expressions for Thermal Contact Conductance . . . 33

3.4.1 Solid Spot Conductance for Fully Plastic Deformation 34

3.4.2 Solid Spot Conductance for Elastic Deformation 35

3.4.3 Variation of Surface Parameters 35

3.5 Effect of Macroscopic Irregularities 36

3.6 Correlations for Solid Spot Conductance 39

3.7 Numerical Example: Solid Spot Conductance 42
3.8 Estimating Contact Parameters and Solid Spot Conductance by Discretization 44
3.8.1 Deformation of a Spherical Asperity 45
3.9 Thermal Boundary Resistance 49
3.9.1 A Simple Model for Total Interface Resistance 50
References 53

4 Gap Conductance at the Interface 55
4.1 Factors Affecting Gas Gap Conductance 55
4.2 The Accommodation Coefficient 58
4.2.1 Effect of Temperature on Accommodation Coefficient 61
4.2.2 Summary of Observations 62
4.3 Effect of Gas Pressure on Gap Conductance 62
4.4 Correlations for Gas Gap Conductance 63
4.5 Numerical Example: Gas Gap Conductance 66
4.6 Recent Research in Gap Conductance 69
4.7 Gas Gap Conductance: Conclusions 72
4.8 Gap Fluid is a Liquid 72
References 75

5 Experimental Aspects 79
5.1 Axial Heat Flow Apparatus 79
5.2 Radial Heat Flow Apparatus 82
5.3 Periodic Contacts 84
5.4 Transient Measurements 85
5.4.1 Transient Methods 85
5.4.2 Transient Measurements: Conclusions 91
5.5 Analogue Methods 91
5.6 Accuracy 94
5.7 Summary 95
References 95

6 Special Configurations and Processes 97
6.1 Bolted or Riveted Joints 97
6.1.1 Stress Distribution at the Bolted Joint Interface 98
6.1.2 Effect of Other Parameters 103
6.1.3 Heat Transfer in Bolted Joints 104
6.1.4 Summary 110
6.2 Cylindrical Joints 110
6.2.1 Numerical Example 115
6.3 Periodic Contacts 119

6.4 Thermal Contact Resistance and Sliding Friction 128
6.4.1 Thermoelastic Instability 131
References 134

7 Control of Thermal Contact Conductance Using Interstitial Materials and Coatings 139
7.1 Solid Interstitial Materials 139
7.2 Metallic Foils 140
7.3 Surface Coatings 144
7.3.1 Constriction Resistance in Plated Contact 144
7.3.2 Thermal Contact Conductance of Coated Surfaces . . . 150
7.3.3 Results for TCC of Coated Surfaces 153
7.3.4 TCC of Coated Surfaces: Recent Works 156
7.4 Insulating Interstitial Materials 159
7.5 Lubricant Films and Greases 159
7.6 Other Interstitial Materials 161
7.6.1 Phase Change Materials 161
7.7 Carbon Nanotubes 162
7.7.1 Carbon Nanotubes: Summary 168
7.8 Other Interstitial Materials: Recent Works 168
7.8.1 Summary: Other Interstitial Materials 175
References 176

8 Major Applications 181
8.1 Contact Heat Transfer in Finned Tubes 181
8.2 Manufacturing Processes 190
8.2.1 Die Casting 190
8.2.2 Injection Moulding 192
8.2.3 Stretch Blow Moulding 195
8.2.4 Rapid Contact Solidification 196
8.2.5 Resistance Spot Welding 198
8.2.6 Thermal Spray Coating 199
8.3 Effective Thermal Conductivity of Packed Beds 200
8.3.1 Correlations for Effective Thermal Conductivity 207
8.3.2 Use of GHP Apparatus to Measure the Effective Conductivity of Packed Beds 209
References 213

9 Additional Topics 217
9.1 Contact Heat Transfer at Very Low Temperatures 217
9.2 Rectification 221
9.2.1 Rectification at Moderate to High Contact Pressures 221
9.2.2 Specimens with Spherical Caps 222

9.2.3 Plane Ended Specimens 224
9.2.4 Microscopic Resistance 226
9.2.5 Rectification at Low Contact Pressures 227
9.2.6 Summary 227
9.3 Hysteresis 228
9.4 Stacks of Laminations 232
9.5 Solid Spot Conductance of Specific Materials 234
9.5.1 Stainless Steel and Aluminium 234
9.5.2 Zircaloy-2/Uranium Dioxide Interfaces 234
9.5.3 Porous Materials 235
9.6 Thermal Contact Resistance in the Presence of Oxide Films 236
9.7 Non–Metallic Materials 238
References 240

10 Concluding Remarks 245
10.1 Control of Thermal Contact Conductance 245
10.1.1 Bare Metallic Junctions 245
10.1.2 Use of Interstitial Materials and Coatings 246
10.1.3 Load Cycling 248
10.1.4 Heat Flow Direction 248
10.1.5 Stacks of Laminations 249
10.2 Recommendations for Further Research 249

Erratum to: Control of Thermal Contact Conductance Using Interstitial Materials and Coatings E1

Author Index 251

Subject Index 257

Nomenclature

The most frequently used symbols in this book are defined below; other symbols are defined in their proper contexts

A	Area
a	Radius of contact spot
B	Correlation distance
b	Radius of the cylinder feeding the contact spot
c	Radius of the contact zone in a bolted joint
C_v	Specific heat at constant volume
CLA	Centre line average
d	Diameter of specimen also plate thickness in bolted joints
E	Modulus of elasticity
E'	Reduced modulus of clasticity

$$= 2\left[\frac{1 - \nu_1^2}{E_1} + \frac{1 - \nu_2^2}{E_2}\right]^{-1}$$

e	Eccentricity also, effectiveness of filler material (Chap. 6)
F	Constriction alleviation factor
f	Degrees of freedom of a gas molecule
g	Temperature jump distance
H	Microhardness
h	Thermal conductance based on unit area
J	Bessel function of the first kind
k	Thermal conductivity
M	Molecular mass
n	Number of contact spots per unit area
N_{KN}	Knudsen number
P	Contact pressure
Q	Rate of heat flow
q	Heat flux
R	Thermal resistance based on unit area
r	Radial co-ordinate

R' Thermal resistance
rms Root mean square
s Mean area of a contact spot
S_u Ultimate compressive strength
T Temperature
TCC Thermal contact conductance
TCR Thermal contact resistance
t Time also, thickness of filler material (Chap. 6)
u Interference
W Load (force)
w Probability density function
x Mass fraction
z Axial co-ordinate

Greek Symbols

α Diffusivity also, coefficient of thermal expansion and accommodation coefficient
δ Mean thickness of air gap also flatness deviation
ε Average clearance
φ Porosity
γ Ratio of specific heats
λ Mean free path also, wavelength
μ Viscosity
ν Poisson's ratio
ρ Radius of curvature
σ Standard deviation (rms value) of asperity heights, also surface tension and stress
ξ Waviness number
ψ Plasticity index

Symbols

c Constriction
cd Disc constriction
eff Effective
g Gas
L Large
m Mean or average
r Real
S Small
s Solid

Chapter 1
Introduction

Microscopic and macroscopic irregularities are present in all practical solid surfaces. Surface roughness is a measure of the microscopic irregularity, whereas the macroscopic errors of form include flatness deviations, waviness and, for cylindrical surfaces, out-of-roundness. Two solid surfaces apparently in contact, therefore, touch only at a few individual spots (Fig. 1.1). Even at relatively high contact pressures of the order of 10 MPa, the actual area of contact for most metallic surfaces is only about 1–2 % of the nominal contact area (see, for example, Bowden and Tabor 1950). Since the heat flow lines are constrained to flow through the sparsely spaced actual contact spots, there exists an additional resistance to heat transfer at a joint. This manifests itself as a sudden temperature drop at the interface.

1.1 Mechanism of Contact Heat Transfer

The heat transfer through a joint may be considered to be made up of three components:

- Conduction through the actual contact spots.
- Conduction through the interstitial medium, such as air.
- Radiation.

The interfacial gap thickness, generally of the order of 1 μm, is too small for convection currents to be set up. Radiation may usually be neglected unless the temperatures at the joint are in excess of 300 °C, unless the temperature drop across the interface is large. Whether the heat transfer by radiation is significant or not depends, therefore, also on the contact resistance. Sometimes it may be justified to consider only the heat conduction through the actual contact spots. However, the area available for heat flow through the interstitial gaps is, frequently, 2–4 orders of magnitude larger than the actual contact area. Hence the heat flow through the gaps cannot be neglected, especially if the solids are relatively poor conductors such as stainless steel and the interface medium is a good conductor.

C. V. Madhusudana, *Thermal Contact Conductance*,
Mechanical Engineering Series, DOI: 10.1007/978-3-319-01276-6_1,

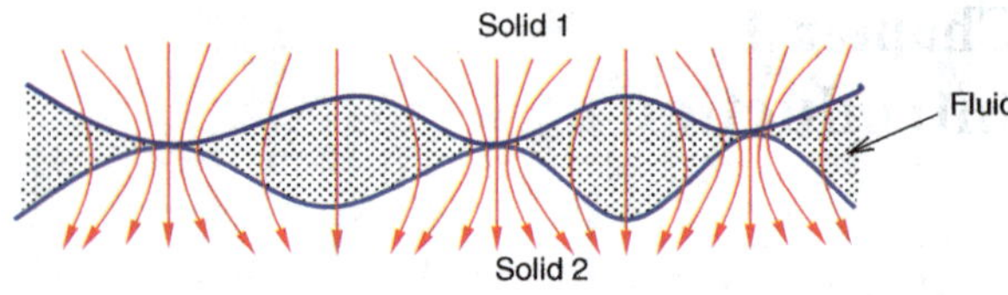

Fig. 1.1 Heat flow through a joint

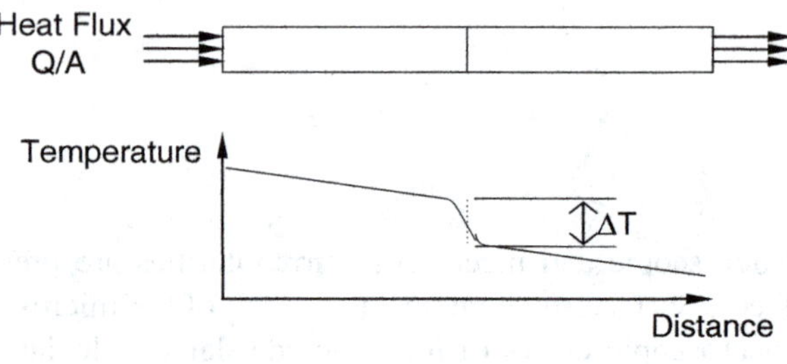

Fig. 1.2 Temperature drop at an interface

Thermal contact conductance, h, is defined as the heat flux to the additional temperature drop, ΔT, due to the presence of the (imperfect) joint (Fig. 1.2):

$$h = \frac{Q/A}{\Delta T} \tag{1.1}$$

In Eq. (1.1), Q is the total heat flow and A is the nominal contact area.

Thermal contact resistance, R, is defined as the reciprocal of thermal contact conductance:

$$R = \frac{A\Delta T}{Q} \tag{1.2}$$

It may be noted that the resistance, as defined above, is usually called specific resistance (or impedance) in heat transfer literature. But the above definition has been adopted by many contact heat transfer researchers, as it is easy to compare the results of different investigators without having to ascertain the area of contact in each case. Another frequently used definition of resistance is based on the total heat flow. This will be designated by the symbol R' in this work:

$$R' = \frac{\Delta T}{Q} \tag{1.3}$$

If it is possible to separate the heat flow through the solid contact spots, Q_s, from the heat flow through the interstitial medium (gas), Q_g, such that

$$Q = Q_s + Q_g \tag{1.4}$$

then the solid spot conductance be defined as:

$$h_s = \frac{Q_s/A}{\Delta T} \tag{1.5}$$

and the gap fluid conductance as:

$$h_g = \frac{Q_g/A}{\Delta T} \tag{1.6}$$

Then, from Eq. (1.1),

$$h = h_s + h_g \tag{1.7}$$

The heat flow through the solid spots, Q_s, is usually determined by conducting the heat transfer tests in a vacuum. The heat flow through the gaps, Q_g, is then determined by conducting the tests in the desired environment and taking the difference in heat flows between the two tests. It will be appreciated, however, that Q_s and Q_g are not independent. The solid spot conductance, as obtained in the vacuum test, will be less than that obtained in a conducting environment; the heat flow lines follow a less tortuous path in the second test. The effect, however, is small unless h_s and h_g are of comparable magnitude.

1.2 Significance of Contact Heat Transfer

Heat transfer systems need to have a high overall efficiency for a number of reasons including energy conservation, limiting maximum operating temperatures and accuracy of temperature measurements. Some applications, where a high value of the thermal contact conductance is necessary, are listed below. Many of these examples are also discussed in detail at the appropriate parts of this book.

1. The fuel/can interface of a nuclear reactor (Dean 1962). It is in the field of nuclear power generating systems that pioneering work on contact heat transfer was carried out in the 1950s and 1960s. The temperature difference between uranium dioxide fuel and the zircaloy sheath could reach several scores of degrees if the contact between them was poor. This could lead to overheating and potential melt down.
2. Bimetallic plain tubes, and attached finned tube types of heat exchangers (Taborek 1987). Because of the different thermophysical properties of the two materials, contact could be reinforced or relaxed, depending on the direction of heat flow. This is one of the factors that dictate the maximum temperature at which the finned tube heat exchanger could operate.
3. Aircraft structural joints subjected to aerodynamic heating (Barzelay et al. 1954). An early impetus to the study of contact heat transfer came from the aircraft industry. In a typical aircraft, the joints are either bolted or riveted leading to imperfect contact between the abutting surfaces.
4. Cooling in electronic systems (Bar-Cohen et al. 2003). With the constant trend toward to micro-miniaturization and consequent increase in power densities, the interface heat transfer in electronic components is an area of great interest.

It may be noted that internal resistances account for approximately 50 % of the overall resistance in a typical heat conduction module.

5. Manufacturing processes including die-casting (Hamasaiid et al. 2010), hot forging (Bourouga et al. 2003), Spray Coating (Heichal and Chandra 2004), resistance spot welding (Lou Lou and Bardon 2001) and injection moulding (Yu et al. 1990: Li 2004). Given below is an example:
A technique called laser-assisted direct imprinting (LADI) uses an excimer laser to irradiate and then heat the silicon surface through a highly transparent quartz mould preloaded onto the silicon (Hsio et al. 2006). The heat transfer problem encountered in the LADI process is the melting/solidification induced by the laser shining through unilaterally transparent binary materials. As a result, thermal contact resistance between the two materials (silicon and quartz) plays an important role in evaluating the temperature distribution, melting duration, and molten depth in this process (Fig. 1.3).

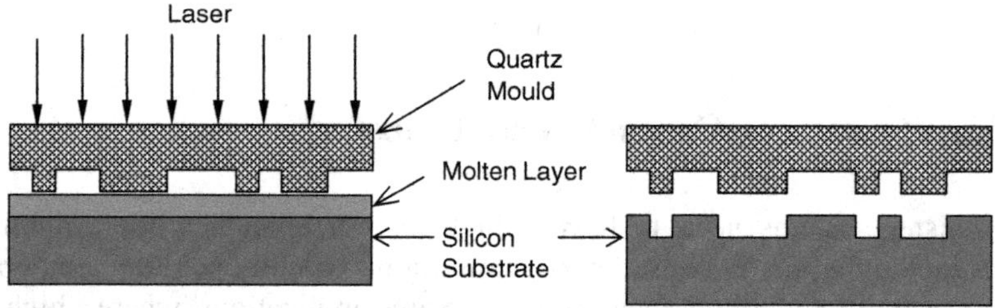

Fig. 1.3 Laser-assisted direct imprinting

6. In sliding contacts (Kennedy 1984; Burton and Burton 1991; Ciavarella and Barber 2005). Apart from the heat loss to the surroundings, the dissipation of heat generated by sliding friction depends on the contact conductance between the two surfaces.
7. Bolted and riveted joints in general (Madhusudana et al. 1988). A dramatic demonstration of how the interface contact heat transfer is improved by the appropriate use of the bolt torque can be seen in Fig. 1.4 drawn on the basis of the experimental results of Yeh et al. (2001). Referring to the figure, the blocks were made of 63.5 × 63.5 mm aluminium alloy (6061 – T6), thickness 50 mm and fastened together by four aluminium bolts with hexagonal nuts. Torque 7.4 Nm; Equivalent Contact Pressure: 3.8 MPa. Heat Flux: 14 kW/m^2.

Other examples where a high value of the TCC is desirable include: the interface between gas turbine blades and rotor, the measurement of surface temperatures by thermocouples, cooking on a hot plate and in reinforced multi-strand overhead electric transmission lines.

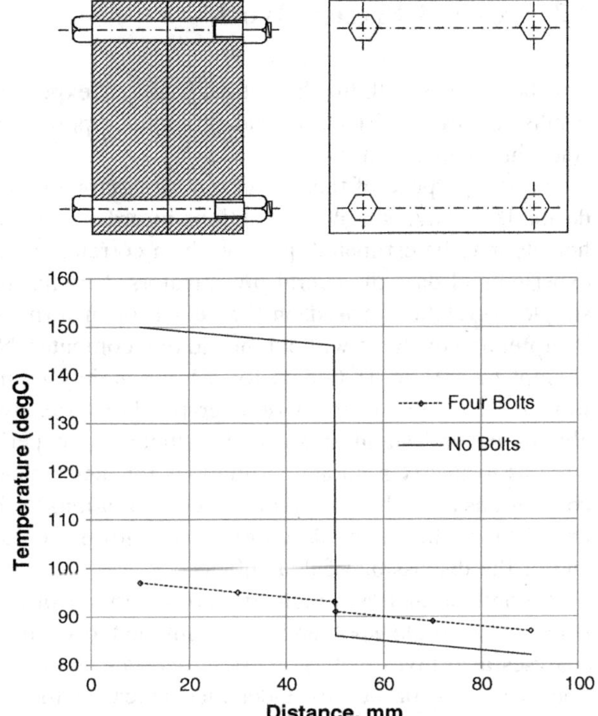

Fig. 1.4 Thermal contact conductance in a bolted joint (based on the data of Yeh et al. 2001)

On the other hand, there are several instances where a high value of TCR, rather than TCC, is desirable. Examples of such applications are:

1. Low conductivity structural supports for the support and storage of cryogenic liquids (Mikesell and Scott 1956). Here the concern is to thermally isolate the liquids from the ambient, as well as the ground supports, which is at a higher temperature.
2. Thermal isolation of spacecraft components that need to be protected from the extremes of temperature associated with space transport (Fletcher 1973).
3. Insulations made of powdered materials (Reiss 1981). Contact resistance plays an important role in the heat transfer between the individual particles of a stationary packed bed. It is one of the major factors affecting the effectiveness of the insulation.

The following reviews list additional applications: Wong (1968), Williams (1968), Madhusudana and Fletcher (1986), Snaith et al. (1986) and Fletcher (1988, 1990).

1.3 Scope of Present Work

This book deals with the theoretical analyses, experimental methods, experimental results, control of TCC, special problems, applications and future directions of study in contact heat transfer.

For the purpose of thermal design of equipment and developing software, it is desirable to have simple correlations in order that the contact heat transfer coefficients may be estimated quickly. Such correlations are usually derived from the experimental data of several investigators. It is not uncommon, however, to find simple correlations based on theoretical open form solutions that are difficult to interpret and evaluate without the aid of a computer. Non-dimensional correlations attempt to present TCC data for a large variety of material combinations in the form of a simple formula or a graph. For extensively used materials such as aluminium and stainless steel, the amount of experimental data available is so large as to justify separate correlations for these materials. There are also special correlations available for particular configurations such as stacks of thin layers and finned tubes. In this book, correlations relevant to each topic are included at the end of the discussion of that subject.

Theoretical analyses include a consideration of the thermal constriction resistance of a single spot and the combined resistance of multiple spots. These analyses also investigate how these resistances may be combined with the variation of surface properties under mechanical or thermal loading in order that the overall thermal resistance of a joint may be estimated. Analyses of the constriction resistance are presented in Chap. 2. The surface and the deformation analyses of the interface are then considered in Chap. 3. These lead to theoretical expressions for the solid spot conductance of a joint consisting of several contact spots as functions of the contact pressure, the surface properties and the mechanical properties of the materials in contact.

The factors influencing the gas gap conductance are described in detail in Chap. 4. These include the surface roughness, the properties of the gas and the gas/solid interface, the gas pressure and the contact pressure. The effects of these factors are then synthesised to yield usable methods for calculating the gap conductance.

In the experimental determination of TCC, the interface is usually the junction of flat surfaces at the ends of two bars (as shown in Fig. 1.2). The geometrical and the mechanical properties of the surfaces are determined prior to the test. The contact pressure is controlled by loading the bars in the axial direction by either mechanical or hydraulic means. In cylindrical joints with radial heat flow, however, the contact pressure is mainly developed as a result of the differential expansion of the cylinders. Therefore a different type of equipment would be needed measure the contact conductance in such a situation. Experimental methods are also available to measure the TCC when the heat flow is transient, when the heat flow is periodic and by analogue. These and other methods will be discussed in Chap. 5.

In bolted joints, the contact resistance is a combination of the macroscopic and the microscopic resistances. In cylindrical joints the contact pressure and contact conductance are inter-dependent. Contact heat transfer in these special configurations is discussed in detail in Chap. 6. The role of contact conductance in periodic contacts and in sliding friction is also reviewed in this chapter.

Traditionally attempts are made to control the TCC by means of different types of interface materials such as foils, films, and wire screens and by coating or plating the contacting surfaces. The performance of these as well as the more recent innovations which make use of carbon nano tubes and phase change materials is assessed in Chap. 7.

Practical applications where TCC is important are visited in detail in Chap. 8. The role of contact heat transfer in finned tube heat exchangers, manufacturing processes and stationary packed beds are discussed in the light of recent research and development in these areas.

Several additional topics in contact heat transfer are treated in Chap. 9. These include: contact heat transfer at low temperatures, heat transfer across stacks of laminations, effect of oxide films, specific materials including non-metallic materials. The effect of heat flow direction on the joint conductance is also considered to see under what conditions rectification can exist. The effect of loading cycles on a joint is reviewed to determine the extent of hysteresis and its practical application.

The final chapter reviews the current state of art in contact heat transfer with particular reference to its control. Recommendations for future work are also included in this chapter.

References

Bar-Cohen A, Watwe AA, Prasher RS (2003) Heat transfer in electronic equipment. In: Bejan A, Kraus AD (eds) Heat transfer handbook. John Wiley, New York, pp 947–1027

Barzelay ME, Tong KN, Hollo C (1954) Thermal conductance of contacts in aircraft joints. US National Advisory Committee for Aeronautics, Washington

Bourouga B, Goizet V, Bardon JP (2003) Predictive model of dynamic thermal contact resistance adapted to the workpiece–die interface during hot forging process. Int Heat Mass Transf 46:565–576

Bowden FP, Tabor D (1950) The friction and lubrication of solids. Oxford University Press, London, pp 20–32

Burton RA, Burton RG (1991) Cooperative interactions of asperities in the thermotribology of sliding contacts. IEEE Transactions, Components and Hybrids, pp 23–35

Ciavarella M, Barber JR (2005) Stability of thermoelastic contact for a rectangular elastic block sliding against a rigid wall. Euro J Mech A (Solids) 24:371–376

Dean RA (1962) Thermal contact conductance between UO_2 and Zircaloy -2, westinghouse Electric corporation, report CVNA-127. Pittsburgh, PA

Fletcher LS (1973) Thermal control materials for spacecraft systems. In: 10th international symposium on space technology and science, Tokyo, pp 579–586

Fletcher LS (1988) Recent developments in contact conductance heat transfer. Trans ASME, J Heat Transf 110:1059–1070

Fletcher LS (1990) A review of thermal enhancement techniques for electronic systems. IEEE Trans Compon Hybrids Manuf Technol 13(4):1012–1021

Hamasaiid A, Dour G, Loulou T, Dargusch MS (2010) A predictive model for the evolution of thermal conductance at the casting–die interfaces in high pressure die casting. Int J Therm Sci 49:365–372

Hsio F-B, Wang D-B, Jen C-P (2006) Numerical investigation of thermal contact resistance between the mold and substrate on laser-assisted imprinting fabrication. Numer Heat Transfer Part A 49:669–682

Kennedy FE Jr (1984) Thermal and thermomechanical effects in dry sliding. Wear 100:453–476

Li H (2004) The role of the thermal contact resistance in injection mold cooling of poly ethylene terephthalate, Ph.D. Thesis, The University of Toledo

Lou Lou T, Bardon JP (2001) Estimation of thermal contact conductance during resistance spot welding. Exp Heat Transf 14:251–264

Madhusudana CV, Fletcher LS (1986) Contact Heat Transfer—The Last Decade. AIAA J 24(3):510–523

Madhusudana CV, Peterson GP, Fletcher LS (1988) Effect of non-uniform pressures on the thermal conductance in bolted and riveted joints. Trans ASME J Energy Resour Technol 112:174–182

Mikesell RP, Scott RB (1956) Heat conduction through insulating supports in very low temperature equipment. J Res Nat Bur Stand (US) 57(6):371–378

Reiss H (1981) An evacuated powder insulation for a high temperature Na/S battery, AIAA Paper 81-1107, New York

Snaith B, Probert SD, O'Callaghan PW (1986) Thermal resistance of pressed contacts. Appl Energy 22:31–84

Taborek J (1987) Bond resistance and design temperatures for high-finned tubes—a reappraisal. Heat Transfer Eng 8(2):26–34

Williams A (1968) Heat transfer across metallic joints, institution of engineering (Australia). Mech Chem Eng Trans 4:247–254

Wong HY (1968) A survey of thermal conductance of metallic contacts, Aeronautical Research Council, CIP No. 973, London

Yeh CL, Wen CY, Chen YF, Yeh SH, Wu CH (2001) An experimental investigation of thermal contact conductance across bolted joints. Exp Therm Fluid Sci 25:349–357

Yu CJ, Sunderland JE, Poli C (1990) Thermal contact resistance in injection molding. Polym Eng Sci 30(24):1599–1606

Chapter 2
Thermal Constriction Resistance

It was seen that the contact interface consists of a number of discrete and small actual contact spots separated by relatively large gaps. These gaps may be evacuated or filled with a conducting medium such as gas. In the first case, all of the heat is constrained to flow through the actual contact spots. If the gaps are filled with a conducting medium, however, some of the heat flow lines are allowed to pass through the gaps, that is, they are less constrained and thus the constriction is alleviated to some extent.

Constriction resistance is a measure of the additional temperature drop associated with a single constriction. Let T_0 be the temperature difference required for the passage of heat at the rate Q through a medium when there is no constriction and T the temperature difference required when a constriction is present, all other things remaining the same. Then the constriction resistance R_c is defined as:

$$R_c = \frac{(T - T_0)}{Q} \tag{2.1}$$

In this chapter, the theory pertaining to the constriction resistance is derived first when the constriction is in isolation, that is, when the effect of the adjacent spots is ignored. Next the constriction resistance of a single spot when it is surrounded by similar spots is determined. The contact conductance is the sum of the conductances of all of the spots existing on the interface. The average radius of these contact spots and their number can be determined by means of surface and deformation analyses so that the conductance may be finally evaluated as a function of the surface parameters, material properties and the contact pressure.

2.1 Circular Disc in Half Space

The logical starting point for the discussion of constriction resistance is to consider the resistance associated with a circular area located on the boundary of a semi-infinite medium. This is equivalent to assuming that:

C. V. Madhusudana, *Thermal Contact Conductance*,
Mechanical Engineering Series, DOI: 10.1007/978-3-319-01276-6_2,

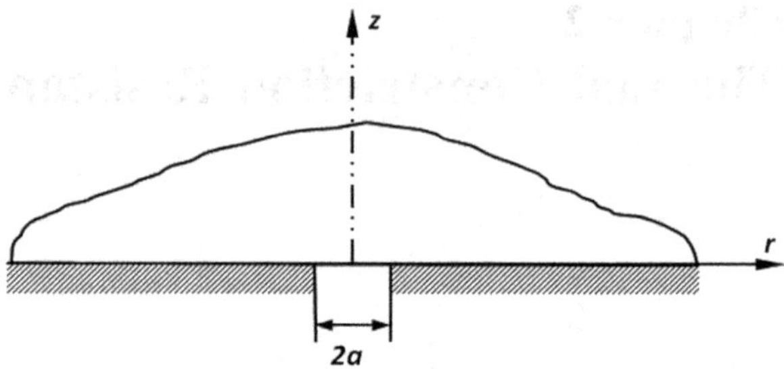

Fig. 2.1 Disc constriction in half-space

a. The constriction is small compared to the other dimensions of the medium in which heat flow occurs.
b. The constriction of heat flow lines is not affected by the presence of other contact spots.
c. There is no conduction of heat through the gap surrounding the contact spot.

The problem is illustrated in Fig. 2.1. Many solutions to this problem are available (see, for example, Llewellyn-Jones 1957; Holm 1957). We will describe here, in some detail, the method used by Carslaw and Jaeger (1959). It is believed that such detail is necessary in order to appreciate fully the mathematical complexities involved in the analytical solutions of even the simplest configurations. In what follows, frequent reference is made to the work of Gradshteyn and Ryzhik (1980). The formulas of this reference will be indicated by G-R followed by the formula number.

The equation of heat conduction in cylindrical co-ordinates, with no heat generation is:

$$\frac{\partial^2 T}{\partial r^2}+\frac{1}{r}\frac{\partial T}{\partial r}+\frac{\partial^2 T}{\partial z^2}=0 \tag{2.2}$$

Using the method of separation of variables, we seek a solution of the form:

$$T(r,z)=R(r)Z(z) \tag{2.3}$$

so that Eq. (2.2) may be written as

$$R''Z+\left(\frac{Z}{r}\right)R'+Z''R=0$$

Dividing through by RZ and separating the variables. We get

$$\frac{\left(R''+\frac{R'}{r}\right)}{R}=-\frac{Z''}{Z}=-\lambda^2$$

Thus Eq. (2.2) is reduced to two ordinary differential equations:

$$\frac{d^2R}{dr^2}+\frac{1}{r}\frac{dR}{dr}+\lambda^2 R=0 \tag{2.4a}$$

and

$$\frac{d^2Z}{dz^2} - \lambda^2 Z = 0 \tag{2.4b}$$

Equation (2.4a) is a form of Bessel's differential equation of order zero and a solution of this is $J_0(\lambda r)$ and a solution of Eq. (2.4b) is $e^{-\lambda z}$. Therefore, Eq. (2.2) is satisfied by $e^{-\lambda z} J_0(\lambda r)$ for any λ. Hence

$$T = \int_0^\infty e^{-\lambda z} J_0(\lambda r) f(\lambda) d\lambda \tag{2.5}$$

will also be a solution if $f(\lambda)$ can be chosen to satisfy the boundary conditions at $z = 0$.

At $z = 0$, the solution (2.5) reduces to

$$T_c = \int_0^\infty J_0(\lambda r) f(\lambda) d\lambda \tag{2.6}$$

In the problem being considered, at $z = 0$, there is no heat flow over the region $r > a$. Also, in the same plane, the region $r < a$ could be at constant temperature or, alternatively, at uniform heat flux. These two cases are considered below.

1. The contact area is maintained at constant temperature T_c over $0 < r < a$

According to G-R 6.693.1,

$$\begin{aligned} Int_\nu &= \int_0^\infty J_\nu(\alpha x) \frac{\sin(\beta x)}{x} dx \\ &= \frac{1}{\nu} \sin\left(\nu \arcsin \frac{\beta}{\alpha}\right) \quad \beta \leq \alpha \\ &= \frac{\alpha^\nu \sin \frac{\nu\pi}{2}}{\nu\left(\beta + \sqrt{\beta^2 - \alpha^2}\right)^\nu} \quad \beta > ;\alpha \end{aligned}$$

Taking the limit as $\nu \to 0$, (applying L'Hopital's Rule), these integrals turn out to be

$$\begin{aligned} Int_0 &= \arcsin \frac{\beta}{\alpha} \quad \beta < \alpha \\ Int_0 &= \frac{\pi}{2} \quad \beta > \alpha \end{aligned}$$

Hence, if we take

$$\begin{aligned} Int_0 &= \arcsin \frac{\beta}{\alpha} \quad \beta \leq \alpha \\ Int_0 &= \frac{\pi}{2} \quad \beta > \alpha \end{aligned}$$

in Eq. (2.6), we get for the temperature at $z = 0$

$$\frac{2T_c}{\pi}\int_0^\infty J_0(\lambda r)\frac{\sin(\lambda a)}{\lambda}d\lambda = \frac{2T_c}{\pi}\arcsin\frac{\lambda}{\lambda} = \frac{2T_c}{\pi}\frac{\pi}{2} = T_c, \quad r \le a$$

Since the temperature is independent of z for $r > a$, this satisfies the other boundary condition, namely, no heat flow over rest of the plane at $z = 0$.

Substituting for $f(\lambda)$ in Eq. (2.5)

$$T = \frac{2T_c}{\pi}\int_0^\infty e^{-\lambda z}J_0(\lambda r)\frac{sin(\lambda a)}{\lambda}d\lambda \tag{2.7}$$

we get, from G-R 6.752.1

$$T = \frac{2T_c}{\pi}\arcsin\frac{2a}{\sqrt{(r+a)^2+z^2}+\sqrt{(r-a)^2+z^2}} \tag{2.8}$$

Note that, for $0 < r \le a$,

$$-\left(\frac{\partial T}{\partial z}\right)_{z=0} = \frac{2T_c}{\pi}\int_0^\infty J_0(\lambda r)\sin(\lambda a)d\lambda = \frac{2T_c}{\pi}\left(\frac{1}{\sqrt{a^2-r^2}}\right) \tag{2.9}$$

from G-R 6.671.7.

The heat flow over the circle $0 < r \le a$,

$$\begin{aligned}
Q &= -2\pi k\int_0^a \left(\frac{\partial T}{\partial z}\right)_{z=0} rdr \\
&= -2\pi k\frac{2T_c}{\pi}\int_0^a \left[\int_0^\infty -\lambda e^{-\lambda z}J_0(\lambda r)\frac{sin(\lambda a)}{\lambda}d\lambda\right]_{z=0} rdr \\
&= 4kT_c\int_0^a \left[\int_0^\infty J_0(\lambda r)\sin(\lambda a)d\lambda\right] rdr \\
&= 4kT_c\int_0^\infty \sin(\lambda a)\left[\int_0^a J_0(\lambda r)rdr\right]d\lambda \\
&= 4kT_c\int_0^\infty \frac{1}{\lambda}\sin(\lambda a)[aJ_1(\lambda a)]d\lambda
\end{aligned}$$

from G-R. 6.561.5. Therefore,

$$\begin{aligned}
Q &= 4akT_c\int_0^\infty \frac{1}{\lambda}\sin(\lambda a)[J_1(\lambda a)]d\lambda \\
&= 4akT_c(1)
\end{aligned}$$

from G-R 6.693.1.

The constriction is finally given by:

$$R_{cd1} = \frac{T_c - 0}{Q} = \frac{1}{4ak} \tag{2.10}$$

2. The contact area is subjected to uniform heat flux
In this case, the boundary conditions at $z = 0$ are:

$$-k\frac{\partial T}{\partial z}\begin{cases} = q, \text{ for } 0 \le r < a \\ = 0, \text{ for } r > a \end{cases} \tag{2.11}$$

where q is the heat flux.
Differentiating Eq. (2.5) with respect to z,

$$\frac{\partial T}{\partial z} = \int_0^\infty -\lambda e^{-\lambda z} J_0(\lambda r) f(\lambda) d\lambda$$

Applying the first of the two boundary conditions (at $z = 0$) in Eq. (2.11),

$$-\frac{\partial T}{\partial z} = \int_0^\infty \lambda J_0(\lambda r) f(\lambda) d\lambda = \frac{q}{k}$$

Considering the integral (G-R, 6.512.3)

$$\int_0^\infty J_0(\lambda r) J_1(\lambda a) d\lambda \begin{cases} = 0 \text{ for } r > a \\ = \dfrac{1}{2a} \text{ for } r = a \\ = \dfrac{1}{a} \text{ for } r < a \end{cases}$$

we see that

$$f(\lambda) = \left(\frac{qa}{k}\right)\frac{J_1(\lambda a)}{\lambda} \tag{2.12}$$

so that the solution is

$$T = \left(\frac{qa}{k}\right)\int_0^\infty e^{-\lambda z} J_0(\lambda r)\frac{J_1(\lambda a)}{\lambda} d\lambda \tag{2.13}$$

The average temperature, T_{av} over $0 < r \le a$ and $z = 0$ is

$$\begin{aligned} T_{av} &= \frac{1}{\pi a^2}\int_0^a T(2\pi r)dr \\ &= \frac{2q}{ak}\int_0^\infty \frac{J_1(\lambda a)}{\lambda}\left\{\int_0^a J_0(\lambda r)dr\right\}d\lambda \end{aligned}$$

From G-R 6.561.5

$$T_{av} = \frac{2q}{ak}\int_0^\infty \frac{J_1(\lambda a)}{\lambda}\left\{a\frac{J_1(\lambda a)}{\lambda}\right\}d\lambda$$

or

$$T_{av} = \frac{2q}{k}\int_0^\infty \left\{\frac{J_1^2(\lambda a)}{\lambda^2}\right\}d\lambda$$

Using the result from G-R 6.574.2 for the integral in the above expression,

$$T_{av} = \frac{8qa}{3\pi k} \tag{2.14}$$

The heat flow rate is

$$Q = \pi a^2 q$$

Hence the constriction resistance for the uniform heat flux condition is

$$R_{cd2} = \frac{T_{av}}{Q} = \frac{8}{3\pi^2 ak} = \frac{0.27}{ka} \tag{2.15}$$

This is about 8 % larger than the constriction resistance R_{cd1}obtained for the uniform temperature condition.

2.2 Resistance of a Constriction Bounded by a Semi-infinite Cylinder

In a real joint there will be several contact spots. Each contact spot of radius a_i may be imagined to be fed by a cylinder of larger radius b_i as shown in Fig. 2.2. Note that the sum of areas of all of the contact spots is equal to the real contact area A_r, while the sum of the cross sectional areas of all of the cylinders is taken as equal to the nominal (apparent) contact area A_n.

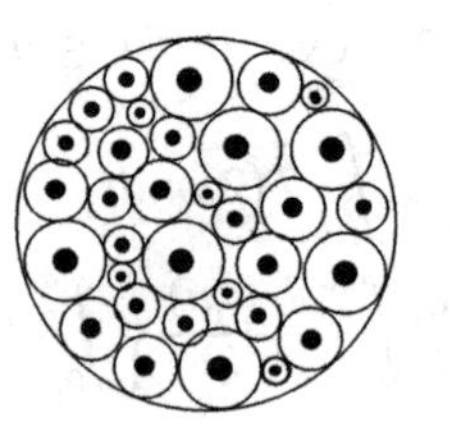

Idealized View of Contact Plane

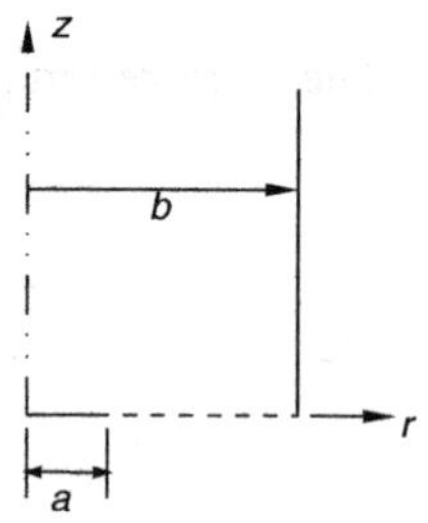

Constriction Bounded by a Semi-Infinite Cylinder

Fig. 2.2 Modelling of a single contact spot in a cluster of spots

2.2.1 Contact Area at Uniform Temperature

It is further assumed that there is no cross flow of heat between the adjacent cylinders. There is also no heat flow across the gap between adjacent contact spots; that is, the contact st is surrounded by a vacuum in the contact plane.

There are several solutions available to this problem. The following analysis is based on the solution described Mikic and Rohsenow (1966) and Cooper et al. (1969).

The boundary conditions defined by the problem are:

$$T = \text{constant};\ z = 0,\ 0 < r \leq a \tag{2.16a}$$

$$-k\frac{\partial T}{\partial z} = 0;\ z = 0,\ r > a \tag{2.16b}$$

$$-k\frac{\partial T}{\partial z} = \frac{Q}{\pi b^2 k};\ z \to \infty \tag{2.16c}$$

$$-k\frac{\partial T}{\partial r} = 0;\ r = b \tag{2.16d}$$

$$-k\frac{\partial T}{\partial r} = 0;\ r = 0 \tag{2.16e}$$

To satisfy the boundary conditions (2.16c) and (2.16e), the solution to Eq. (2.2) should be in the form:

$$T = \frac{Q}{\pi b^2 k} z + \sum_{n=1}^{\infty} C_n e^{-\lambda z} J_0(\lambda_n r) + T_0 \tag{2.17}$$

From the boundary condition in Eq. (2.16d), we get

$$J_1(\lambda_n b) = 0 \tag{2.18}$$

Here b = 3.83171, 7.01559, 10.17347, etc. (Abramovitz and Stegun 1968a). Also by integrating Eq. (2.17) over the *whole of the interfacial area* $(0 < r < b)$ at $z = 0$ and using Eq. (2.18), we see that the average temperature for this area is T_0.

In Eq. (2.17), the C_n's are to be determined from the boundary conditions Eqs. (2.16a) and (2.16b) at $z = 0$. However, these boundary conditions are mixed. To overcome this problem, the Dirichlet boundary condition in Eq. (2.16a) is replaced by a heat flux distribution for the circular disc in half space [see Eq. (2.9)]:

$$-k\frac{\partial T}{\partial z} = \frac{Q}{2\pi a\sqrt{a^2 - r^2}};\ z = 0,\ 0 < r \leq a \tag{2.19}$$

This approximation will lead to a nearly constant temperature distribution over the area specified by $0 < r \leq a$, especially for small values of ε, where $\varepsilon = (a/b)$.

However, from Eq. (2.17), at $z = 0$,

$$-k\frac{\partial T}{\partial z} = k\left[\frac{Q}{\pi b^2 k} + \sum_{n=1}^{\infty} C_n \lambda_n J_0(\lambda_n r)\right] \tag{2.20}$$

From (2.19) and (2.20), therefore

$$\left[\frac{Q}{\pi b^2 k} + \sum_{n=1}^{\infty} C_n \lambda_n J_0(\lambda_n r)\right] = \frac{Q}{2\pi ak\sqrt{a^2 - r^2}}$$

To utilize the orthogonality property of the Bessel function both sides of the above equation are multiplied by $rJ_0(\lambda_n r)$ and integrated over the appropriate ranges to yield

$$\frac{Q}{\pi b^2 k}\int_0^b rJ_0(\lambda_n r)dr + C_n\lambda_n\int_0^b rJ_0^2(\lambda_n r)dr = \frac{Q}{2\pi ak}\int_0^a \frac{rJ_0(\lambda_n r)}{\sqrt{a^2 - r^2}}dr$$

(see G-R 6.521.1).
However

$$\int_0^b rJ_0(\lambda_n r)dr = \frac{b}{\lambda_n}J_1(\lambda_n b)$$

This is equal to zero by virtue of Eq. (2.18).
From the orthogonality property (Abramovitz and Stegun 1968b)

$$\int_0^b rJ_0^2(\lambda_n r)dr = \frac{b^2}{2}J_0^2(\lambda_n b)$$

and

$$\int_0^a \frac{rJ_0(\lambda_n r)}{\sqrt{a^2 - r^2}}dr = \frac{sin(\lambda_n a)}{\lambda_n}$$

(see G-R 6.554.2).
Therefore

$$C_n = \left(\frac{Q}{\pi ka}\right)\frac{\sin(\lambda_n a)}{(\lambda_n b)^2 J_0^2(\lambda_n b)}$$

Substituting for C_n in Eq. (2.17)

$$T = \frac{Q}{\pi b^2 k}z + \left(\frac{Q}{\pi ka}\right)\sum_{n=1}^{\infty}\frac{e^{-\lambda z}\sin(\lambda_n a)J_0(\lambda_n r)}{(\lambda_n b)^2 J_0^2(\lambda_n b)} + T_0 \tag{2.21}$$

The mean temperature over the interface ($z = 0$) is then obtained by

$$T_m = \frac{1}{\pi a^2}\int_0^a \left(\frac{Q}{\pi k a}\right)\sum_{n=1}^{\infty}\frac{e^{-\lambda z}\sin(\lambda_n a)J_0(\lambda_n r)}{(\lambda_n b)^2 J_0^2(\lambda_n b)}2\pi r dr + \frac{1}{\pi b^2}\int_0^b T_0 2\pi r dr$$
$$= \frac{Q}{4ka}\left(\frac{8}{\pi a^2}\right)\sum_{n=1}^{\infty}\frac{\sin(\lambda_n a)}{(\lambda_n b)^2 J_0^2(\lambda_n b)}\int_0^a rJ_0(\lambda_n r)dr + T_0$$

This results in

$$T_m = \frac{Q}{4ka}\left(\frac{8}{\pi}\right)\left(\frac{b}{a}\right)\sum_{n=1}^{\infty}\frac{sin(\lambda_n a)J_1(\lambda_n a)}{(\lambda_n b)^3 J_0^2(\lambda_n b)} + T_0 \tag{2.22}$$

The factor 1/(4 ka) in the above expression represents the disc constriction resistance of Eq. (2.10). The thermal resistance between $z = 0$ and $z = L$ (for large L) is given by

$$R_t = \frac{T_m - T_{z=L}}{Q} = \frac{T_m}{Q} + \frac{L}{\pi k b^2} - \frac{T_0}{Q}$$

Hence the *additional* resistance due to constriction is

$$R = R_t - \frac{L}{\pi k b^2} = \frac{T_m - T_0}{Q}$$

Substituting for T_m from Eq. (2.22)

$$R = \frac{1}{4ka}\left(\frac{8}{\pi}\right)\left(\frac{b}{a}\right)\sum_{n=1}^{\infty}\frac{\sin\left[\lambda_n b\left(\frac{a}{b}\right)\right]J_1\left[\lambda_n b\left(\frac{a}{b}\right)\right]}{(\lambda_n b)^3 J_0^2(\lambda_n b)} = R_{cd1}F\left(\frac{a}{b}\right) \tag{2.23}$$

in which

$$F\left(\frac{a}{b}\right) = \left(\frac{8}{\pi}\right)\left(\frac{b}{a}\right)\sum_{n=1}^{\infty}\frac{\sin\left[\lambda_n b\left(\frac{a}{b}\right)\right]J_1\left[\lambda_n b\left(\frac{a}{b}\right)\right]}{(\lambda_n b)^3 J_0^2(\lambda_n b)} \tag{2.24}$$

is called the *constriction alleviation factor*.

Yovanovich (1975) obtained expressions for the constriction alleviation factor for heat flux functions of the form

$$\left[1 - \frac{r^2}{a^2}\right]^m; \; z = 0, \; 0 < r \le a$$

Hid results for $m = -0.5$ were identical to that of Mikic, as expected.

Other solutions to the above problem include those of Roess (as presented by Weills and Ryder (1949)), Hunter and Williams (1969), Gibson (1976), Rosenfeld and Timsit (1981) and Negus and Yovanovich (1984). The algebraic expressions derived for the constriction alleviation factor by Roess, Gibson, and Negus and Yovanovich are somewhat similar

Table 2.1 Comparison of constriction alleviation factors

a/b	Roess (Eq. 2.25)	Mikic (Eq. 2.24)	Gibson (Eq. 2.26)	N-Y (Eq. 2.27)
0.1	0.8594	0.8584	0.8594	0.8594
0.2	0.7205	0.7202	0.7209	0.7208
0.3	0.5853	0.5851	0.5865	0.5865
0.4	0.4558	0.4557	0.4586	0.4586
0.5	0.3340	0.3341	0.3398	0.3395
0.6	0.2230	0.2231	0.2328	0.2318

$$F_{\text{Roess}} = 1 - 1.4093(a/b) + 0.2959(a/b)^3 + 0.0524(a/b)^5 + \cdots \tag{2.25}$$

$$F_{\text{Gibson}} = 1 - 1.4092(a/b) + 0.3380(a/b)^3 + 0.0679(a/b)^5 + \cdots \tag{2.26}$$

$$F_{\text{Negus-Yovanovich}} = 1 - 1.4098(a/b) + 0.3441(a/b)^3 + 0.0435(a/b)^5 + \cdots \tag{2.27}$$

The constriction alleviation factors obtained by Eqs. (2.24)–(2.27) are compared in Table 2.1. The first 120 terms were used in evaluating the series in Eq. (2.24).

2.2.2 Contact Area Subjected to Uniform Heat Flux

In this case, the boundary condition, represented by Eq. (2.16a) is replaced by

$$-k\frac{\partial T}{\partial z} = \text{constant}; \quad z = 0,\ 0 < r \leq a \tag{2.28}$$

The constriction alleviation factor for this problem was theoretically derived by Yovanovich (1976):

Note: 1. In the following expression $\varepsilon = a/b$

2. The constriction resistance is non-dimensionalized by multiplying it by $k\sqrt{A_c}$, that is, by $ka\sqrt{\pi}$ and not $4ak$

$$F'(\varepsilon) = \frac{16}{\pi\varepsilon}\sum_{n=1}^{\infty}\frac{J_1^2(\lambda_n\varepsilon)}{\lambda_n^3 J_0^2(\lambda_n)} \tag{2.29}$$

Negus, Yovanovich and Beck (1989) provided the following correlation to evaluate $F'(\varepsilon)$

$$F'(\varepsilon) = 0.47890 - 0.62498\varepsilon + 0.11789\varepsilon^3 + \cdots \tag{2.30}$$

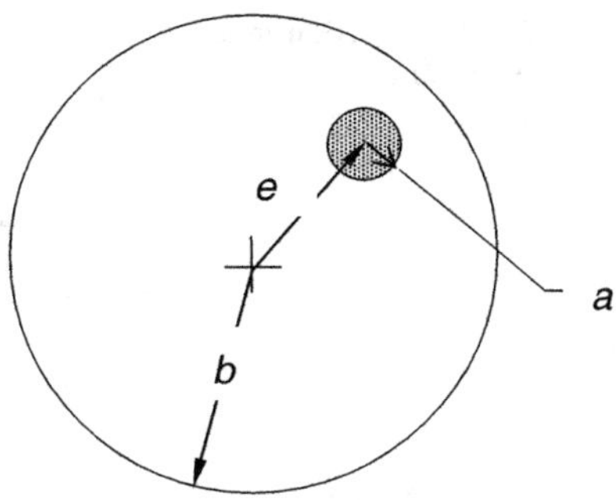

Fig. 2.3 Eccentric constriction

2.3 Eccentric Constrictions

In Sect. 2.2, the contact spot was assumed to be concentric with the feeding flux tube. The subject of eccentric constrictions has been studied by Cooper et al. (1969), Sexl and Burkhard (1969) and others. A more recent work is by Bairi and Laraqi (2004) who presented an analytical solution to calculate the thermal constriction resistance for an eccentric circular spot with *uniform flux* on a semi-infinite circular heat flux tube. This solution is developed using the finite cosine Fourier transform and the finite Hankel transform (Fig. 2.3).

The authors proposed a dimensionless correlation to calculate the constriction alleviation factor as a function of ε and the eccentricity *e*:

$$\Psi^* = \frac{\Psi}{\Psi_0} = 1 + \left[1.5816\left({}^{a}\!/_{b}\right)^{0.0528} - 1\right]\left(\frac{e}{b-a}\right)^{1.76}\left(\frac{a}{b-e}\right)^{0.88} \tag{2.31}$$

In this equation the authors took the following correlation for Ψ_0, the constriction factor for zero eccentricity (Negus et al. 1989):

$$\begin{aligned}\Psi_0 = {} & 0.47890 - 0.62076(a/b) \\ & + 0.114412(a/b)^3 + 0.01924(a/b)^5 + 0.00776(a/b)^7\end{aligned} \tag{A}$$

However, the expression given in (A) is the *constriction factor for a circular cross section at the end of a square tube!* The correct factor that the authors should have used is in Eq. (2.30). Therefore the accuracy of Eq. (2.31) is open to question.

2.4 Constriction in a Fluid Environment

In this case, the boundary conditions at the contact plane (z = 0) are as shown in Fig. 2.4 in which k_f is the thermal conductivity of the fluid (gas) and δ is the effective gap thickness.

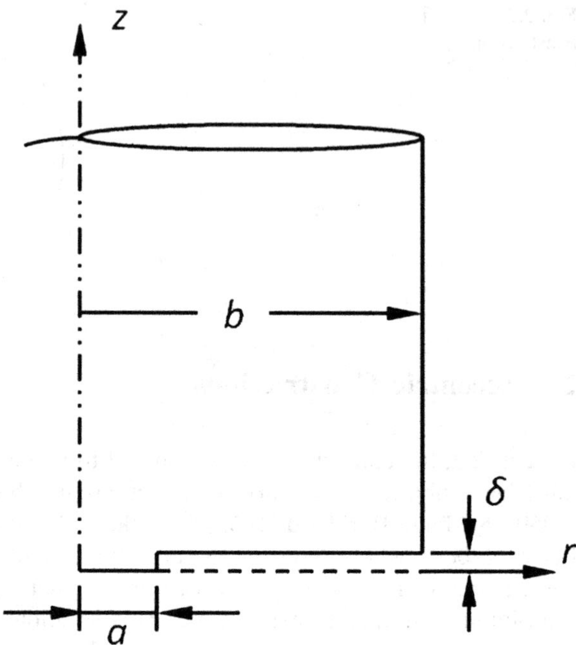

Fig. 2.4 Constriction in a fluid environment

Approximate solutions to the above problem have been obtained by Cetinkale and Fishenden (1951), Mikic and Rohsenow (1966) and Tsukizoe and Hisakado (1972). A later work by Das and Sadhal (1998) presents an analytical solution to the problem of a constriction surrounded by an interstitial fluid. An 'exact' solution was presented by Sanokawa (1968), but the results of this work were not in a readily usable form. In any case, the model used in these analyses, as illustrated above in Fig. 2.4, is somewhat artificial—the gap thickness is abruptly changed from zero thickness to a finite thickness at $r = a$. The thickness is expected to increase gradually. Any analytical solution is, therefore, likely to be complicated and a digital computer would be still required to evaluate the results. For this reason, a numerical solution is perhaps more suitable for the solution of this type of problems.

In a large number of situations, the heat flow through the gas gap is small compared to the heat flow through the solid contact spots. In such cases, the fluid conductance may be estimated by dividing the fluid conductivity by the effective gap thickness. This may then be added to the solid spot conductance to obtain the total conductance. Factors affecting the gas gap conductance are discussed in detail in Chap. 4.

2.5 Constrictions of Other Types

Apart from the solutions discussed above, problems pertaining to constrictions of other shapes and boundary conditions have been analysed by various researchers. These are listed in Table 2.2.

Table 2.2 Constriction resistance—representative works

No.	Reference	Configuration	Approach
1	Mikic and Rohsenow (1966)	Strip of contact and rectangles in vacuum	Analytical
2	Yip and Venart (1968)	Single and multiple constrictions in vacuum	Analogue
3	Veziroglu and Chandra (1969)	Two dimensional, symmetric and eccentric constrictions in vacuum	Analytical and analogue
4	Williams (1975)	Conical constrictions in vacuum	Experimental
5	Yovanovich 1976	Circular annular constriction at the end of a semi-infinite cylinder in vacuum	Analytical
6	Gibson and Bush (1977)	Disc constriction in half space in conducting medium	Analytical
7	Major and Williams (1977)	Conical constrictions in vacuum	Analogue
8	Schneider (1978)	Rectangular and annular contacts in vacuum half space	Numerical
9	Yovanovich et al. (1979)	Doubly connected areas bounded by circles, squares and triangles in vacuum	Analytical
10	Madhusudana (1979a, b), (1980)	Conical constrictions at the end of a long cylinder, in vacuum and in conducting medium	Numerical and experimental
11	Major (1980)	Conical constrictions in vacuum	Numerical
12	Negus et al. (1988)	Circular contact on coated surfaces in vacuum	Analytical
13	Das and Sadhal (1992)	Two dimensional gaps at the interface of two semi-infinite solids in a conducting environment	Analytical
14	Madhusudana and Chen (1994)	Annular constriction at the end of a semi-infinite cylinder in vacuum	Analytical and analogue
15	Olsen et al. (2001a, b) (2002)	Coated conical constrictions in vacuum, gas and with radiation	Numerical

References

Abramovitz M, Stegun IA (1968a) Handbook of mathematical functions. Dover, New York, p 409

Abramovitz M, Stegun IA (1968b) Handbook of mathematical functions. Dover, New York, p 48

Bairi A, Laraqi N (2004) The thermal constriction resistance for an eccentric spot on a circular heat flux tube. Trans ASME J Heat Transf 128:652–655

Carslaw HS, Jaeger JC (1959) Conduction of heat in solids, 2nd edn. Clarendon Press, Oxford, pp 214–217

Cetinkale TN, Fishenden M (1951) Thermal conductance of metal surfaces in contact. In: Proceedings of the *general* discussion on *heat* transfer, Institute of Mechanical Engineers, London, pp 271–275

Cooper MG, Mikic BB, Yovanovich MM (1969) Thermal contact conductance. Int J Heat Mass Transf 12:279–300

Das AK, Sadhal SS (1992) The effect of interstitial fluid on thermal constriction resistance. Trans ASME J Heat Transf 114:1045–1048

Das AK, Sadhal SS (1998) Analytical solution for constriction resistance with interstitial fluid in the gap. Heat Mass Transf 34:111–119

Gibson RD (1976) The contact resistance for a semi-infinite cylinder in vacuum. Appl Energy 2:57–65

Gibson RD, Bush AW (1977) The flow of heat between bodies in gas-filled contact. Appl Energy 3:189–195

Gradshteyn IS, Ryzhik M (1980) Tables of integrals, series and products. Academic Press, New York

Holm R (1957) Electric contacts, theory and application, 4th edn. Springer, New York, pp 11–16

Hunter A, Williams A (1969) Heat flow across metallic joints—the constriction alleviation factor. Int J Heat Mass Transf 12:524–526

Llewellyn-Jones F (1957) The physics of electric contacts. Oxford University Press, New York, pp 13–15

Madhusudana CV, Chen PYP (1994) Heat flow through concentric annular constrictions. In: Proceedings of the 10th international heat transfer conference, paper 3-Nt 18, Institution of Chemical Engineers, Rugby

Madhusudana CV (1979a) Heat flow through conical constrictions in vacuum and in conducting media. In: AIAA 14th thermophysics conference, paper 79-1071, Orlando, Florida

Madhusudana CV (1979b) Heat flow through conical constrictions. AIAA J 18:1261–1262

Madhusudana, CV (1980) Heat flow through conical constrictions, AIAA Journal, 18: 1261–1262

Major SJ, Williams A (1977) The solution of a steady conduction heat transfer problem using an electrolytic tank analogue, Inst Engrs (Australia). Mech Eng Trans 7–11

Major, SJ (1980) The finite difference solution of conduction problems in cylindrical co-ordinates, IE (Aust), Mech Eng Trans, Paper M 1049

Mikic BB, Rohsenow WM (1966) Thermal contact resistance, Mechenical Engineering Department report no. DSR 74542-41, MIT, Cambridge

Negus KJ, Yovanovich MM (1984) Constriction resistance of circular flux tubes with mixed boundary conditions by superposition of neumann solutions. In: ASME Paper 84-HT-84, American Society of Mechanical Engineers

Negus KJ, Yovanovich MM, Thompson JC (1988) Constriction resistance of circular contacts on coated surfaces:effect of boundary conditions. J Thermophys Heat Transf 12(2):158–164

Negus KJ, Yovanovich MM, Beck JV (1989) On the nondimentionalization of constriction resistance for semi-infinite heat flux tubes. Trans ASME J Heat Transf 111:804–807

Olsen E, Garimella SV, Madhusudana CV (2001a) Modeling of constriction resistance at coated joints in a gas environment. In: 2nd international symposium on advances in computational heat transfer, Palm Cove, Qld, Australia, 20–25 May 2001

Olsen E, Garimella SV, Madhusudana CV (2001b) Modeling of constriction resistance at coated joints with radiation. In: 35th national heat transfer conference, Anaheim, California, 10 June 2001

Olsen EL, Garimella SV, Madhusudana CV (2002) Modeling of constriction resistance in coated joints. J Thermophys Heat Transf 16(2):207–216

Rosenfeld AM, Timsit RS (1981) The potential distribution in a constricted cylinder: an exact solution. Q Appl Maths 39:405–417

Sanokawa K (1968) Heat transfer between metallic surfaces. Bull JSME 11:253–293

Schneider, G.E, (1978), Thermal constriction resistances due to arbitrary contacts on a half space—numerical solution. In: AIAA Paper 78-870, Amer Inst Aeronautics and Astronautics, New York

Sexl RU, Burkhard DG (1969) An exact solution for thermal conduction through a two dimensional eccentric constriction. Prog Astro Aero 21:617–620

Tsukizoe T, Hisakado T (1972) On the mechanism of heat transfer between metal surfaces in contact—Part 1, heat transfer. Jpn Res 1(1):104–112

Veziroglu TN, Chandra S (1969) Thermal conductance of two dimensional constrictions. Prog Astro Aero 21:591–615

Weills ND, Ryder EA (1949) Thermal resistance of joints formed by stationary metal surfaces. Trans ASME 71:259–267

Williams A (1975) Heat flow through single points of metallic contacts of simple shapes. Prog Astro Aero 39:129–142

Yip FC, Venart JES (1968) Surface topography effects in the estimation of thermal and electrical contact resistance. Proc I Mech Eng 182(3):81–91

Yovanovich, M.M. (1975), General expressions for constriction resistance due to arbitrary flux distributions, AIAA Paper 75-188, Amer Inst Aeronautics and Astronautics, New York

Yovanovich, MM (1976) General thermal constriction parameter for annular contacts on circular flux tubes, AIAA Journal, 14:822–824

Yovanovich MM, Martin KA, Schneider GE (1979) Constriction resistance of doubly-connected contact areas under uniform heat flux, AIAA Paper 79-1070, The American Institute of Aeronautics and Astronautics

Chapter 3
Solid Spot Thermal Conductance of a Joint

A real joint consists of a numerous contact spots and each spot is associated with two resistances—constriction and spreading—in series. The thermal contact conductance (or resistance) is a combination of individual conductances (or resistances) of all of the spots in the joint. In this chapter, the number and the average size of contact spots a given pair of surfaces in contact are established by statistical analysis of the contacting surfaces. These are then related to the mechanical properties of the contacting materials by deformation analyses. The thermal, the surface and the deformation analysis are then combined to obtain a relationship between the conductance the contact pressure. This chapter also includes brief discussions of macroscopic errors of form eccentric contacts the thermal boundary resistance.

3.1 Multiple Spot Contact Conductance

When both sides of the contact spot are considered (Fig. 3.1), the total resistance is simply the sum of the resistances for each side of the contact. Therefore, if k_1 and k_2 are the thermal conductivities of the two solids, then the resistance associated with a single contact spot is given by:

$$R = \frac{F}{4ak_1} + \frac{F}{4ak_2} = \frac{F}{2ak} \tag{3.1}$$

where F is the constriction alleviation factor as defined in Chap. 2 and

$$k = \frac{2k_1k_2}{k_1 + k_2} \tag{3.2}$$

is the harmonic mean of the two conductivities.

The thermal conductance of a single spot is, therefore:

$$h = \frac{2ak}{F} \tag{3.3}$$

C. V. Madhusudana, *Thermal Contact Conductance*,
Mechanical Engineering Series, DOI: 10.1007/978-3-319-01276-6_3,

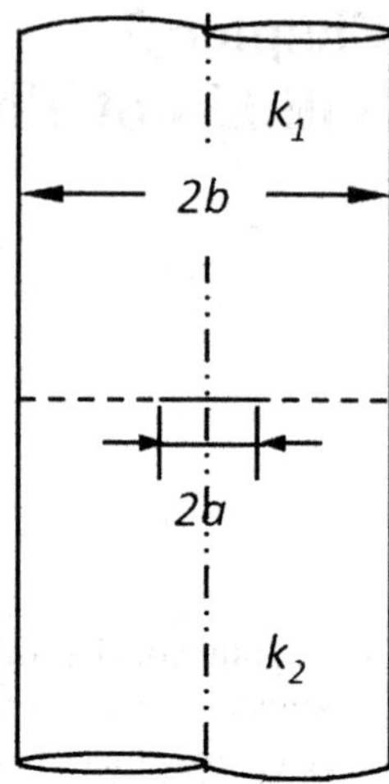

Fig. 3.1 Two cylinders contacting over a central contact spot

Hence, if there are n contact spots in the interface, the thermal conductance of the solid spots is:

$$h_s = 2k\sum_n \frac{a_i}{F_i} \tag{3.4}$$

Neglecting the variation in F_i and writing

$$\sum_n a_i = na_m \tag{3.5}$$

where a_m is the average radius of the contact spots, we obtain

$$h_s = \frac{2na_m k}{F} \tag{3.6}$$

It now remains to determine n and a_m by means of surface and deformation analyses.

Note Neglecting the variation in F_i amounts to assuming a *uniform* distribution of contact spots. The theoretical investigation of Das and Sadhal (1999) on the constriction resistance of randomly distributed contacts showed that the resistance will be higher in the latter situation. They suggested that this may be due to the high probability of finding larger segments of insulated areas, due to some degree of clustering, in the random arrangement. This conclusion was confirmed by Laraqi (2003) who noted that “the constriction resistance of random contacts is systematically larger than the one of the regular contacts having the same number of contacts and surface (area) of spots” Laraqi also noted that the constriction resistance increased when there was a large difference in the minimum and the maximum contact radii.

3.2 Estimation of the Number and Average Size of the Contact Spots

Several authors have described the method of calculating the number and the size of contact spots when two rough surfaces are placed in contact under specified mechanical pressure (see, for example, Tsukizoe and Hisakado 1965, 1968; Mikic and Rohsenow 1966; Greenwood 1967; Greenwood and Tripp 1970).

The following analysis is based on the method presented by Kimura (1970). In the analysis, it is assumed that the heights and slopes of the surface profiles are randomly distributed. It is also assumed that the probability densities of the height and the slope are independent. Other assumptions relate to particular cases and will be stated in their proper contexts.

We first consider the contact of a rough surface with a flat smooth surface. Let $z = f(x.y)$ be the profile height above the x–y plane defined by

$$\iint_A z dx dy = 0$$

where A is the nominal contact area. This indicates that the x–y plane is located at the mean height of the profile.

The profile slopes in the x- and the y- directions are given by

$$z_x = \frac{\partial z}{\partial x}$$

and

$$z_y = \frac{\partial z}{\partial y}$$

Let $w(z)$ be the probability density function so that $w(z)dz$ is the probability that z is in the range from z to $z + dz$. Then, if A_r is the real area of contact and ε is the average clearance (see Fig. 3.2) at a given pressure P

$$\frac{A_r}{A} = prob(z > \varepsilon) = \int_{\varepsilon}^{\infty} w(z)dz \tag{3.7}$$

Consider a *unit length of surface profile* parallel to the x-axis. In this length, let n_x be the number of sections in which the profile is in the real area of contact.

The joint probability that z is in the range $z + dz$ and z_x is in the range z_x to $z_x + dz_x$ is given by:

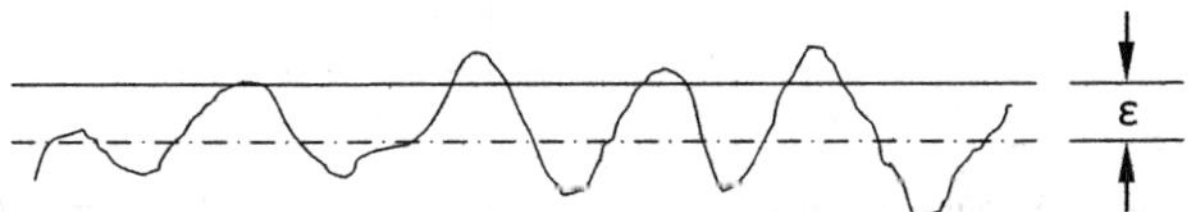

Fig. 3.2 Clearance between a smooth flat plane and a rough surface

$$w(z, z_x)dzdz_x$$

Note that this is also the fraction of length per unit length, in the x-direction, that the height and the slope are in the specified ranges.

But to cross the interval dz, a length $dz/|z_x|$ is needed in the x- direction (Fig. 3.3). Hence the expected number of crossings, per unit length, at z = ε and $z_x = z_x$ is given by

$$\frac{w(\varepsilon, z_x)dzdz_x}{\frac{dz}{|z_x|}} = |z_x|w(\varepsilon, z_x)dz_x$$

Now z_x can have any value at $z = \varepsilon$. Hence the total number of crossings at $z = \varepsilon$ is

$$\int_{-\infty}^{\infty} |z_x|w(\varepsilon, z_x)dz_x$$

In the above integration, both upward and downward crossings are included. The number of *sections* will be half the above value. Further, since the probability densities of the slope and the height are independent, the joint probability is simply the product of the two probabilities. Therefore

$$n_x = \frac{1}{2}w(\varepsilon)\int_{-\infty}^{\infty} |z_x|w(z_x)dz_x \tag{3.8}$$

The number of sections n_y, per unit length in the y-direction may be found similarly.

Let a_x and a_y be the mean lengths of the contact spots in the x- and y- directions, respectively (Fig. 3.4). Then

$$a_x n_x = a_y n_y = \int_{\varepsilon}^{\infty} w(z)dz = \frac{A_r}{A} \tag{3.9}$$

from Eq. (3.7).

If s is the mean area of a contact spot, then a non-dimensional shape factor g may be defined as

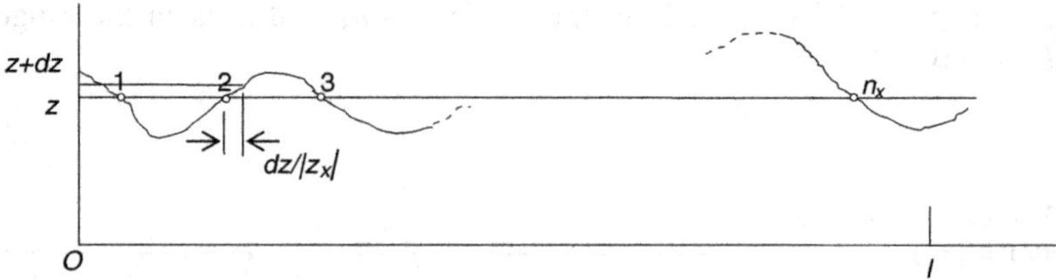

Fig. 3.3 Number of crossings between z and $z + dz$ (after Kimura 1970)

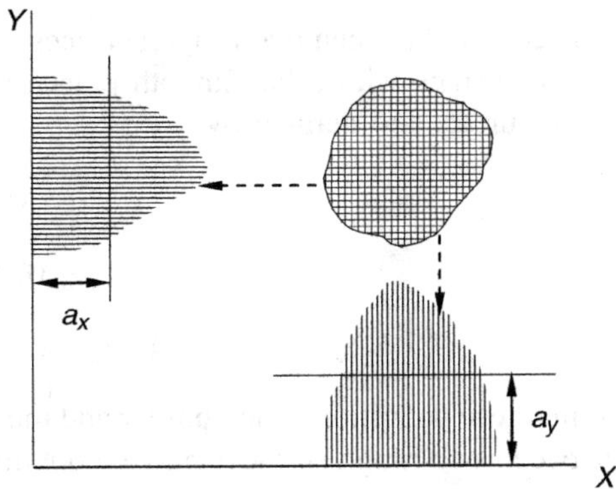

Fig. 3.4 Mean lengths of contact spots

$$g = \frac{s}{a_x a_y} \tag{3.10}$$

Note that g = 1 for a rectangular and g = 4/π and an elliptic cross-section respectively. If n is the number of contact spots per unit area, then

$$ns = \frac{A_r}{A} \tag{3.11}$$

Therefore,

$$n = \frac{A_r}{A}\left(\frac{1}{s}\right) = \frac{A_r}{A}\left(\frac{1}{g a_x a_y}\right) = \left(\frac{A_r}{A}\right)\left(\frac{1}{g}\right)\frac{n_x n_y}{\left(\frac{A_r}{A}\right)^2}$$

using Eq. (3.9) in the last step.

Hence

$$n = \left(\frac{1}{g}\right)\left(\frac{A}{A_r}\right) n_x n_y$$

Substituting for n_x from Eq. (3.8), and a similar one for n_y, the *number of contact spots per unit area* is obtained as

$$n = \left(\frac{1}{4g}\right)\left(\frac{A}{A_r}\right)\{w(\varepsilon)\}^2 \int_{-\infty}^{\infty} |z_x| w(z_x) dz_x \int_{-\infty}^{\infty} |z_y| w(z_y) dz_y \tag{3.12}$$

Then, from Eq. (3.11), the *average area of a contact spot* is

$$s = \frac{4g\left(\frac{A_r}{A}\right)^2}{\{w(\varepsilon)\}^2 \int_{-\infty}^{\infty} |z_x| w(z_x) dz_x \int_{-\infty}^{\infty} |z_y| w(z_y) dz_y} \tag{3.13}$$

The contact between two rough surfaces may be treated as that of an "equivalent" rough surface with a flat smooth plane. The heights and the slopes of the equivalent surface are defined by

$$z_e = z_1 + z_2$$

$$(z_x)_e = (z_x)_1 + (z_x)_2$$

$$(z_y)_e = (z_y)_1 + (z_y)_2$$

With these definitions, the number and the average area of contact spots for contact between two rough surfaces may be obtained if the slopes and heights are taken to be the equivalent ones in Eqs. (3.12) and (3.13).

3.2.1 Gaussian Distribution of Heights and Slopes

Here

$$w(z) = \frac{1}{\sqrt{2\sigma\pi}} exp\left(-\frac{z^2}{2\sigma^2}\right) \tag{3.14}$$

in which σ is the standard deviation of profile heights. Similar expressions for $w(z_x)$ and $w(z_y)$ may be written by replacing σ by the standard deviations, σ_x and σ_y, for the profile slopes.

Then the average clearance is given by

$$\left(\frac{A_r}{A}\right) = \frac{1}{2} erfc\left(\frac{\varepsilon}{\sigma\sqrt{2}}\right), \varepsilon \geq 0 \tag{3.15}$$

Note that the above equation is applicable for $(A_r/A) < 0.5$. ε would be negative if (A_r/A) is greater than 0.5.

Solving for the clearance,

$$\varepsilon = \sigma\sqrt{2} erfc^{-1}\left(\frac{2A_r}{A}\right) \tag{3.16}$$

From Eqs. (3.12) and (3.13),

$$n = \left(\frac{1}{4\pi^2 g}\right)\frac{\sigma_x\sigma_y}{\sigma^2}\left(\frac{A}{A_r}\right) exp\left[-2\left\{erfc^{-1}\left(\frac{2A_r}{A}\right)\right\}^2\right] \tag{3.17}$$

$$s = \left(4\pi^2 g\right)\frac{\sigma^2}{\sigma_x\sigma_y}\left(\frac{A_r}{A}\right)^2 exp\left[2\left\{erfc^{-1}\left(\frac{2A_r}{A}\right)\right\}^2\right] \tag{3.18}$$

In Eqs. (3.15)–(3.18),

$$erfc(x) = 1 - erf(x)$$

is the complementary error function; $erfc^{-1}(x)$ is the inverse of $erfc(x)$.

For Gaussian distribution, the standard deviation of profile heights for the equivalent surface is given by

$$\sigma_e^2 = \sigma_1^2 + \sigma_2^2 \tag{3.19}$$

and by similar equations for the profile slopes.

3.3 Deformation Analysis

The objective of this section is to present the means of estimating (A_r/A) which is required to evaluate the number and the average area of contact spots using Eqs. (3.17) and (3.18). In order to do this, we have to first determine whether the deformation will be plastic or elastic in a given situation. The concept of a "plasticity index" is useful in this connection.

3.3.1 The Plasticity Index

The deformation of an asperity will be elastic up to some given load above which some plastic flow will occur. If w_p is the compliance of the asperity at the *onset of plastic flow*, then the deformation will be entirely elastic for compliances less than this value.

For a sphere in contact with a smooth plane, this compliance is given by (Tabor 1951)

$$w_p = r\left(\frac{H}{E'}\right)^2 \tag{3.20}$$

in which r is the radius of the spherical asperities, H is the hardness of the asperity material, and E' is the reduced elastic modulus for the two materials in contact.

If σ is the standard deviation of the asperity heights, then the value of w_p/σ could be used as the criterion for whether the contact is elastic, plastic, or lies within the load-dependent range. For Gaussian distribution of profile heights, Greenwood (1967), proposed a plasticity index:

$$\psi_g = \left(\frac{E'}{H}\right)\sqrt{\frac{\sigma}{r}} \tag{3.21}$$

Greenwood showed that the surfaces with the plasticity index values greater than 1, representing many freshly made surfaces, will have plastic contact at the lightest loads. Values below 0.7, which can be obtained by careful polishing, give elastic contact even at heavy loads.

The random surface model of Whitehouse and Archard (1970), also results in a somewhat similar plasticity index (Tabor, 1975):

$$\psi_A = \left(\frac{E'}{H}\right)\sqrt{\frac{\sigma}{B}} \tag{3.22}$$

in which B is the correlation distance corresponding roughly to the spacing between asperities of equal heights.

Tabor (1975) demonstrated that in each case,

a. Plastic deformation occurred when the plasticity index reached the value of 1.
b. The coefficient of (E'/H) represented the mean slope of the asperities.

From the second of these observations it follows that:

$$\psi = \left(\frac{E'}{H}\right)\tan\theta \tag{3.22a}$$

Mikic (1974) also demonstrated that it is the slope of the asperities that controls the mode of deformation for a given pair of materials in contact.

Another version of the plasticity index using the first three moments of the power spectral density of the surface profile, rather than the radius of curvature and the standard deviation, has been proposed by Bush and Gibson (1979).

3.3.2 Ratio of Real to Apparent Area of Contact

In the following discussion, P is the load (mechanical pressure) between the contact surfaces.

(i) Purely Plastic Deformation
Here it is assumed that all asperities in contact are deforming at the same contact pressure, H, which is the microhardness of the softer of the two materials. Hence

$$A_r H = AP$$

or

$$\frac{A_r}{A} = \frac{P}{H} \tag{3.23}$$

In the above equation, it is implicitly assumed that the displaced material just disappears; that is, the conservation of volume is not observed. This will not cause serious errors for low contact pressures (large separations). For larger loads, Mikic

(1974) proposed that the above equation be modified as:

$$\frac{A_r}{A} = \frac{P}{H + P} \tag{3.24}$$

Both Eqs. (3.23) and (3.24) assume there is no work hardening of the surfaces in contact. They also assume that the hardness is constant, but microhardness does depend on indentation size (or applied load) as observed by many researchers. See, for example, Li et al. (1998) who also noted that the hardness depends on the method of surface preparation as well.

(ii) Elastic Deformation

For elastic deformations, the contact area can be related to the displacement and, hence, the load by Hertzian theory. (Note that the Hertzian theory does not apply to non-metals such as plastics—these will be discussed at a later section). For spherically shaped asperities whose heights follow a Gaussian distribution, Roca and Mikic (1971) proved that

$$\frac{A_r}{A} = \frac{P\sqrt{2}}{E'\tan\theta} \tag{3.25}$$

where $\tan\theta$ is the mean absolute profile of the slope.

Note that, even for elastic deformation, the real area of contact is proportional to the contact pressure P.

3.4 Theoretical Expressions for Thermal Contact Conductance

We are now in a position to combine the results of the thermal, the surface, and the deformation analyses so as to derive usable expressions for the solid spot thermal conductance.

For circular contact spots, the shape factor, g, in Eq. (3.10) is $4/\pi$ and s $= \pi a_m^2$ so that $a_m^2 = s/\pi$. Hence from Eq. (3.18)

$$a_m^2 = \left(\frac{1}{\pi}\right)(4\pi^2)\left(\frac{4}{\pi}\right)\left(\frac{\sigma}{\tan\theta}\right)^2\left(\frac{A_r}{A}\right)^2 exp\left[2\left\{erfc^{-1}\left(\frac{2A_r}{A}\right)\right\}^2\right] \tag{3.26}$$

where $\tan^2\theta = \sigma_x\sigma_y$. Hence

$$a_m = 4\left(\frac{\sigma}{\tan\theta}\right)\left(\frac{A_r}{A}\right)exp\left[\left\{erfc^{-1}\left(\frac{2A_r}{A}\right)\right\}^2\right] \tag{3.27}$$

Similarly, the Eq. (3.17) for the number of contact spots may be written as

$$n = \left(\frac{1}{16\pi}\right)\left(\frac{\tan\theta}{\sigma}\right)^2\left(\frac{A}{A_r}\right)exp\left[-2\left\{erfc^{-1}\left(\frac{2A_r}{A}\right)\right\}^2\right] \tag{3.28}$$

Multiplying Eqs. (3.27) and (3.28),

$$na_m = \left(\frac{1}{4\pi}\right)\left(\frac{\tan\theta}{\sigma}\right)exp\left[-\left\{erfc^{-1}\left(\frac{2A_r}{A}\right)\right\}^2\right] \tag{3.29}$$

Substituting for na_m in Eq. (3.6), we obtain

$$h_s = \left(\frac{1}{2\pi}\right)\left(\frac{k\tan\theta}{\sigma F}\right)exp[-X] \tag{3.30}$$

where

$$X = \left\{erfc^{-1}\left(\frac{2A_r}{A}\right)\right\}^2 \tag{3.31}$$

It may be noted that until this point, no assumption has been made regarding the mode of deformation.

3.4.1 Solid Spot Conductance for Fully Plastic Deformation

Here we use Eq. (3.23) calculating the true contact area so that

$$X = \left\{erfc^{-1}\left(\frac{2P}{H}\right)\right\}^2 \tag{3.32}$$

Hence the solid spot conductance can be calculated for any contact pressure P. The analysis of Mikic (1974) yielded a similar result

$$h_s = 1.13\left(\frac{k\tan\theta}{\sigma}\right)\left(\frac{P}{H}\right)^{0.94} \tag{3.33}$$

However, note that in Eq. (3.33), $\tan\theta$ is the mean absolute slope of the profiles. In Eq. (3.30), $\tan\theta$ is the standard deviation of the slope distribution.

If the displaced volume is taken into account, then Eq. (3.24) is used to determine the area ratio in Eq. (3.32). Similarly (P/H) is replaced by $[P/(P + H)]$in Eq. (3.33).

3.4.2 Solid Spot Conductance for Elastic Deformation

The value of (A_r/A), in this case, is given by Eq. (3.25). The equation derived by Mikic for h_s when the asperities deform elastically is

$$h_s = 1.55\left(\frac{k\tan\theta}{\sigma}\right)\left(\frac{P\sqrt{2}}{E'\tan\theta}\right)^{0.94} \tag{3.34}$$

Here the numerical coefficient is different from 1.13 of Eq. (3.33). This may be explained by observing the fact that, for a given separation, the contact area in an elastic deformation is twice that in purely plastic deformation. This implies that for a given area ratio, the sum of the contact radii, in elastic deformation, is $\sqrt{2}$ times the value in plastic deformation. Also note that the elastic conductance is only a weak function of $\tan\theta$ indicating that the variations in $\tan\theta$ will not significantly alter the conductance.

3.4.3 Variation of Surface Parameters

It has long been recognised that the magnitudes of the surface parameters, as measured by a profilometer such as the Talysurf, are dependent on the sampling length (also known as cut-off length). Figure 3.5 shows the variation of roughness, slope and radius of curvature of the asperity with the sampling length. Data for the graph refer to bead-blasted stainless steel specimens and is taken from Li et al. (1998).

It can be seen that while the roughness remains substantially constant, there is a slight variation in slope with the sampling length. The variation in radius of curvature is more marked. Similar trends for roughness and slope were noted by Ju and Farris (1996). They plotted the variation of curvature (reciprocal of radius

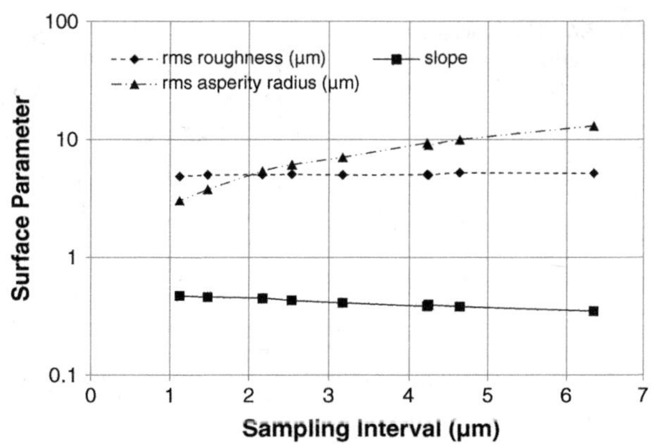

Fig. 3.5 Variation of surface parameters with the sampling interval

of curvature) which showed the opposite trend to that of the radius of curvature, as it should.

In view of the above observations, it is recommended that the sampling length be also reported when reporting the surface parameters, although this is not of particular significance when only the roughness is reported.

3.5 Effect of Macroscopic Irregularities

The analyses presented so far assume that the surfaces are rough, but flat. In practice, many manufactured surfaces possess some degree of deviation from flatness. It is also likely that some waviness would be present as a result of the machining process, such as grinding and turning, by which the surfaces were produced. It is necessary to correct the expression for the conductance (or resistance) to account for such departure from flatness.

Clausing and Chao (1965) suggested that the flatness deviations may be accounted for by means of the "spherical cap" model (Fig. 3.6). The apparent contact area is then divided into two regions.

a. The non contact region, which contains few or no microscopic contact areas.
b. The contact region, where the density of microcontacts are high.

The flow of heat is first constrained to the large scale contact areas and then further constrained to the microscopic contacts within the macroscopic areas. The total resistance R_t is given by

$$R_t = R_L + R_s \tag{3.35}$$

in which R_L and R_s represent the macrscopic and the microscopic resistances, respectively. A film resistance may also need to be added for oxidised surfaces.

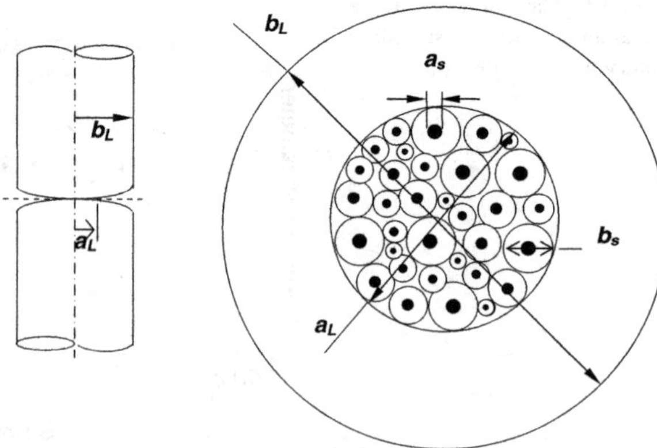

Fig. 3.6 The macroscopic constriction

Since the resistance is the reciprocal of the conductance, we get, using Eq. (3.6),

$$R_t = \frac{F_L}{2a_L k} + \frac{F_s}{2a_s k} \tag{3.36a}$$

As an example, consider the contact of two identical cylinders, each of radius b_L and flatness deviation δ. Then, from geometry, the radius of curvature of the spherical caps is given by

$$\rho \approx \frac{b_L^2}{2\delta}$$

By Hertzian equation for the elastic deformation of spheres, the radius of the macroscopic contact area for a contact force W is (Timoshenko and Goodier 1970) is

$$a_L = 1.109\left(\frac{W\rho}{2E}\right)^{\frac{1}{3}} = 1.109\left(\frac{Wb_L^2}{4\delta E}\right)^{\frac{1}{3}} \tag{3.36b}$$

In this expression, E is the elastic modulus and the Poisson's ratio is taken to be 0.3. Substituting for a_L in Eq. (3.36a), we see that the macroscopic resistance varies as $W^{(-1/3)}$.

Yovanovich (1969) demonstrated that the theory of Clausing and Chao may be extended to predict the resistance of rough, wavy surfaces. In this case, the macroscopic contact areas, called "contour areas". are formed due to the waviness of contacting surfaces.

Thomas and Sayles (1975) considered that the vertical section through a surface contained a continuous spectrum of wavelengths (Fig. 3.7). The largest wavelengths with the largest amplitudes correspond to large scale errors of form; the shorter and smaller wavelengths constitute the waviness; the shortest and smallest wavelengths represent the roughness. Therefore, waviness and roughness should be discussed in terms of bandwidths of this spectrum rather than that of fixed wavelengths.

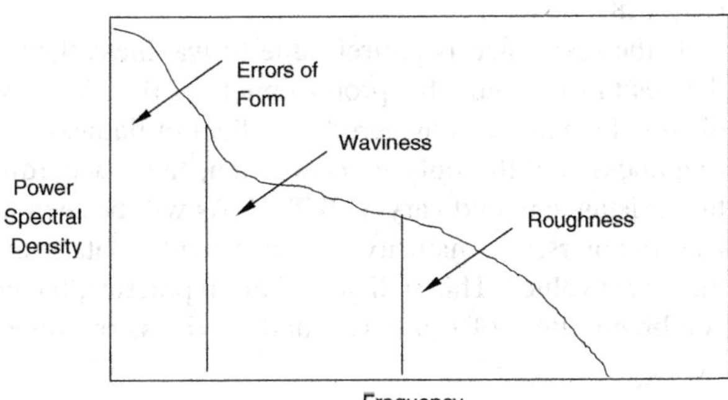

Fig. 3.7 Power Spectrum of surface profile (after Thomas and Sayles 1975)

In their analysis, it was assumed that:

a. One of the contacting surfaces is smooth and flat, whereas the other surface is randomly smooth and isotropically rough.
b. The cut-off wavelength is 2d where d is the diameter of the specimen.
c. Initially, the surfaces touch each other at three points (the summits), so that there will be three macroscopic contact areas.
d. All of the summits have the same height and the same radius of curvature, r_0.
e. The radius of curvature of the contour area is given by Hertzian relation for the elastic contact area between a sphere of radius r_0 and a flat.

The analysis showed that

$$r_0 = 4.2 \times 10^{-3} \frac{d^2}{\sigma} \tag{3.37}$$

Hence the contour area radius was shown to be given by

$$\frac{a_L}{b_L} = 0.44\zeta^{1/3} \tag{3.38}$$

where

$$\zeta = \frac{W}{E' d\sigma} \tag{3.39}$$

is the waviness number and W is the load (force).

It was also shown that σ was proportional to $d^{1/2}$. Hence, if σ_m was the measured roughness at a cut-off length of L, then

$$\sigma = \sigma_m \left(\frac{d}{L}\right)^{1/2} \tag{3.40}$$

From Eq. (3.40), when $\zeta = 1$, the value of (a_L/b_L) is 0.44, and using a constriction factor such as that given by Eq. (2.25), we can see that the macroscopic resistance is less than half the value of the disc constriction resistance. Hence Thomas and Sayles suggested that the effect of waviness may be neglected for $\zeta > 1$. For higher values of ζ, the effect of roughness only needs to be considered (Fig. 3.8).

If the resistance is entirely due to waviness then, from Eqs. (3.38) and (3.39), the resitance would be proportional to $W^{-1/3}$. This is the same result that we obtained when we considered the effect of flatness deviation. On the other hand, if roughness was the only consideration, then, according to Eqs. (3.33) and (3.34), the resistance would vary as $W^{-0.94}$. As will be seen in the next section, and also in later chapters, the majority of experimental data show an index that is in between these two values. This indicates that, in practice, the contact resistance is due to the combined effect of roughness and waviness or some deviation from flatness.

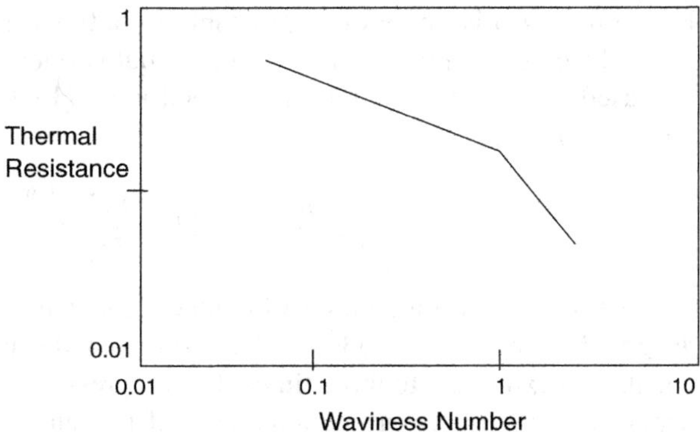

Fig. 3.8 Thermal resistance and the waviness number (after Thomas and Sayles 1975)

3.6 Correlations for Solid Spot Conductance

Over the past fifty years, several different correlations have been proposed for the estimation of solid spot conductance. In what follows, only a few representative correlations will be discussed.

In general, experimental results for solid spot conductance are obtained from test conducted under vacuum conditions. It is clear, from the theoretical analyses in the preceding sections, that the parameters of significance to solid spot conductance should include:

a. The harmonic mean of the thermal conductivities of the two solids
b. The contact pressure
c. The surface roughness
d. The "flow pressure" or similar property of the softer of the two materials
e. The mean junction temperature.

Ideally, the correlation should also allow for the effects of flatness deviation and the mean slope of the surface profile. In the early days of contact heat transfer research, these quantities were not readily measurable and many of the correlations do not take them into account. The mean junction temperature also does not appear in many correlations, although it may be accounted for by the use of values for the thermophysical physical properties appropriate to the temperature.

An early correlation for solid spot thermal conductance was by Laming (1961):

$$\frac{h_s}{k} = \frac{2}{1-F}\left(\frac{1}{\pi\lambda_1\lambda_2}\right)\left(\frac{P}{H}\right)^{0.5} \tag{3.41}$$

in which F is the Roess Constriction alleviation factor.

It should be noted that the above formula was based on the theoretical and experimental results for joints composed of crossed ridges of wavelengths λ_1 and λ_2, respectively, and hence, *the number of contact spots remained constant with load.* For randomly rough surfaces, the number of contact spots increases with load.

Equation (3.41) is, therefore, applicable to only such joints as those composed of crossed wedges or pyramids contacting a flat surface.

Based on experimental data, Mal'kov (1970) proposed the following correlation:

$$\frac{h_s\bar{a}}{k} = 0.118\left(\frac{PC_1}{3S_u}\right)^{0.66} \tag{3.42}$$

Here $\bar{a}$ is the average radius of contact spots which was taken by Mal'kov to be 40 μm. Use was also made of the result that the flow pressure (hardness) was equal, approximately to three times the compressive strength, S_u. C_1 is a factor that depends on the micro-projections, H_{av}, of the surfaces as given by the following relations.

$$C_1 = 1, \text{ for } H_{av1} + H_{av2} > 30\,\mu\text{m}$$

$$C_1 = \left(\frac{30}{H_{av1} + H_{av2}}\right)^{1/3} \text{ for } H_{av1} + H_{av2} < 30\,\mu\text{m}$$

$$C_1 = \left(\frac{15}{H_{av1} + H_{av2}}\right) \text{ for } H_{av1} + H_{av2} < 10\,\mu\text{m}$$

The average heights of micro-projections for different kinds of machining are shown in Table 3.1.

The correlation, indicated by line II in Fig. 3.9, appeared to agree well with most of the 92 experimental data points generated by seven different investigators. It was, however, seen that at high contact pressures the data seemed to be better approximated by line I which has a slope of 0.86. Although the effect of temperature was not directly included, it was suggested that correlations such as this should not be applied to contact zone temperatures exceeding 0.3 times the melting point (in °C) of the (softer) material.

The following correlation proposed by Tien (1968) was based on the theoretical and experimental work of eight sources dealing with nominally flat surfaces:

Table 3.1 Average height of micro-projections

Material	Machining method	Surface finish	H_{av} (μm)
Stainless steel	Turned	∇_4	23.5
Stainless steel	Turned	∇_5	14.0
Stainless steel	Turned	∇_8	2.4
Stainless steel	Ground	∇_8	2.2
Stainless steel	Ground	∇_9	1.2
Stainless steel	Turned and lapped	∇_{10}	0.6 – 0.8
Molybdenum	Ground	∇_9	1.0
Molybdenum	Ground	∇_9	1.07

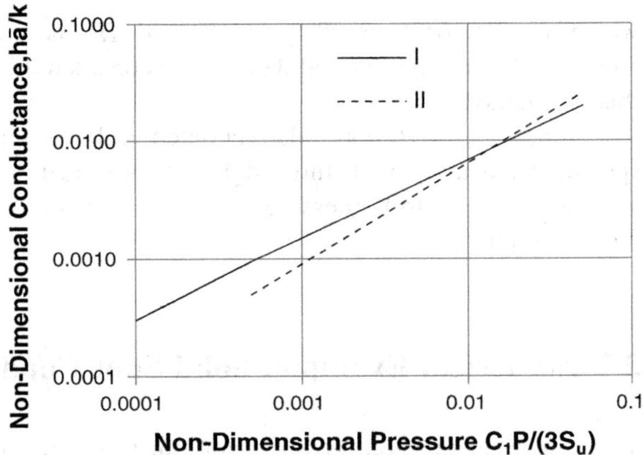

Fig. 3.9 Mal'kov's correlation of experimental data

$$\frac{h_s \sigma}{k} = 0.55 \tan\theta \left(\frac{P}{H}\right)^{0.85} \tag{3.43}$$

The similarity between this equation, derived using dimensional analysis, and Mikic's theoretical derivation, Eq. (3.33), is remarkable.

Fletcher and Gyorog (1971) derived a correlation which accounted for the mean junction temperature. They used results of their own tests as well as those of nine other independent investigators. Elastic modulus E, rather than the hardness, was used to non-dimensionalise the pressure. The units used were not SI—for example, h was expressed in BTU/(hrft^2 °F) and the roughness in micro-inches. They obtained an index equal to 0.56 for the pressure. The correlations produced by Popov (1976) and Antonetti et al. (1993) returned an index of about 0.95 for the contact pressure.

In summary, the correlations for solid spot conductance seem to fall into two categories; those in which the exponent of the pressure is in the range 0.66–0.75 and those for which the exponent is about 0.95. The results for nominally flat contacts seem to fit into the latter class. On the other hand, correlations which collate the works of a large number of investigators generally yield much lower exponents reflecting that practical surfaces contain waviness, flatness deviation or other macroscopic errors of form. The recent work of Bahrami et al. (2004) collates a large number of experimental data and acknowledges the fact that the total resistance is the sum of the microscopic and macroscopic resistances. They proposed the following correlation for the joint resistance R_j in (K/W)

$$R_j^* = 2k_s L R_j = \frac{0.36}{P*} + \frac{L(1 - a_L/b_L)^{1.5}}{a_L} \tag{3.44}$$

where $L = b_L^2/(\sigma/\tan\theta), P^* = P/H^*$ and H^* is the (variable) microhardness. About 83 % of experimental data points considered by them fell within ±15 % of this correlation.

Lambert et al. (2006) also collated a large number of experimental data specifically dealing with three light alloys used in aerospace applications. Their work also takes into flatness deviation. Their results were presented as a series of design graphs.

3.7 Numerical Example: Solid Spot Conductance

The numerical example given below illustrates the steps in determining the solid spot thermal conductance for a joint formed between two nominally flat surfaces. Table 3.2 lists the data used in the example.

Table 3.2 Material property data used in calculations

Property	Aluminium alloy	Stainless steel
Thermal conductivity, k [W/(mK)]	200	16.5
Hardness, H [MPa]	1400	3800
Modulus of elasticity, E [MPa]	$70 \times (10^3)$	$70 \times (10^3)$
Poisson's ratio, ν	0.33	0.29

Calculations will be performed for three separate surface combinations—rough/rough, smooth/smooth and rough/smooth as listed in Table 3.3.

Table 3.3 Surface properties and combinations used in the example

Combination	Aluminium alloy		Stainless steel	
	CLA roughness (μm)	Slope (rad)	CLA roughness (μm)	Slope (rad)
A	1	0.18	1	0.18
B	0.1	0.03	0.1	0.03
C	1	0.18	0.1	0.03

Note The CLA (centre line average or arithmetic average) values for the roughness are given. For Gaussian surfaces rms roughness $\sigma \approx 1.25$ x CLA roughness

The effective properties are calculated as follows:
Effective modulus of elasticity

$$E = 2\left[\frac{1-\nu_1^2}{E_1} + \frac{1-\nu_2^2}{E_2}\right]^{-1} = 114 \times 10^3 \text{ MPa}$$

Effective thermal conductivity

$$k = \frac{2k_1k_2}{k_1 + k_2} = 30.48\,\mathrm{W/(mK)}$$

Effective (or combined) roughness:

$$\sigma = 1.25\left(\sigma_1^2 + \sigma_1^2\right)^{0.5}$$

(See Note under Table 3.3).
Thus

$$\sigma_A = 1.25\left(1^2 + 1^2\right)^{0.5} = 1.77\,\mu\mathrm{m} = 1.77 \times (10)^{-6}\,\mathrm{m}$$

Similarly

$$\sigma_B = 0.177 \times (10)^{-6}\mathrm{m},\ and\ \sigma_C = 1.256 \times (10)^{-6}\mathrm{m}$$

The effective slope is

$$\tan\theta \approx \left\{(\mathrm{slope})_1^2 + (\mathrm{slope})_2^2\right\}^{0.5}$$

Thus

$$\tan\theta_A = 0.254;\ \tan\theta_B = 0.0424;\ \tan\theta_c = 0.182.$$

The plasticity index, as defined in Eq. (3.22a), may be calculated as

$$\psi = \left(\frac{E'}{H}\right)\tan\theta$$

where H is the hardness of the softer material. It is 1400 MPa in this example. Thus:

$$\psi_A = 20.7;\ \psi_B = 3.45;\ \psi_C = 14.82.$$

The plasticity index is greater than 1 for all three combinations indicating that the formula, Eq. (3.33), for plastic deformation is applicable:

$$h_s = 1.13\left(\frac{k\tan\theta}{\sigma}\right)\left(\frac{P}{H}\right)^{0.94}$$

Substituting $k = 30.48\,\mathrm{W/(mK)}$ and $H = 1400\,\mathrm{MPa}$,, we get

$$h_s = 0.038\left(\frac{\tan\theta}{\sigma}\right)P^{0.94}$$

Thus the conductance versus pressure relationships for the three combinations would be:

a. (Rough/Rough): $h_s = 5453P^{0.94}$
b. (Smooth/Smooth): $h_s - 9102P^{0.94}$

Table 3.4 Results of solid spot conductance calculations

Contact pressure (MPa)	Solid spot conductance, W/(m^{2}K)		
	Rough/rough	Smooth/smooth	Rough/smooth
0.1	626	1045	632
0.5	2842	4744	2870
1	5453	9102	5506
5	24755	41321	24996
10	47494	79275	47955
50	215609	359889	217705
100	413652	690457	417673

c. (Rough/Smooth): $h_s = 5506P^{0.94}$

Note

1. The results for conductance will be in W/(m^{2}K) if the pressure is expressed in *MPa.* The results are shown in Table 3.4.
2. The solid spot conductance would be, in fact, the joint thermal conductance in vacuum if radiation is not significant. For heat transfer in non-vacuum conditions, the gas gap conductance needs to be added, as will be explained in the next chapter.
3. It is clear that little enhancement in conductance can be obtained if only one of the surfaces is made smooth. For significant enhancement of conductance the roughness of both surfaces needs to be reduced.

3.8 Estimating Contact Parameters and Solid Spot Conductance by Discretization

A review of the preceding sections reveals that:

- the assumption of an appropriate distribution is necessary to compute the surface parameters
- the hardness is assumed to be constant, although it is known to depend on the contact size
- a decision has to be made whether or not macroscopic resistance, due to flatness deviation for example, is significant.

It is possible to remove some of these assumptions and limitations by discretization of the peak height distribution. Note that while the profile distribution may be Gaussian, the peak heights may not be so distributed (Hunter 1972). The following discussion is based on the work of Li et al. (1998, 2000).

This method depends on first determining whether a particular asperity will deform elastically, plastically or in the intermediate range. Some background information is necessary in this connection.

3.8.1 Deformation of a Spherical Asperity

Sneddon's (1965) solution of the Boussinesq problem yielded the following relation between the deflection and the contact radius for the *elastic* deformation of a flat surface by a sphere:

$$\delta = \frac{1}{2} a \times ln\left(\frac{R+a}{R-a}\right)$$

Now

$$ln\left(\frac{R+a}{R-a}\right) = ln\left(\frac{1+\frac{a}{R}}{1-\frac{a}{R}}\right) = 2\left(\frac{a}{R} + \frac{a^3}{3R^3} + \cdots\right)$$

Since (a/R) is small, the higher order terms in the parentheses may be neglected. Therefore

$$\delta = \frac{1}{2} a \times 2\left(\frac{a}{R}\right)$$

or

$$a^2 = \delta R$$

This is equivalent to assuming that the elastic deflection is divided equally above and below the contact circle (Field and Swain 1993), see Fig. 3.10 below:

When a sphere contacts a flat surface and the deformation is elastic, Hertz's equations (see, for example, Zhan 2001) for the approach and the contact radius are:

$$\delta = \frac{\pi a p_0}{2E^*}$$

and

$$a = \frac{\pi R p_0}{2E^*}$$

where p_0 is the maximum pressure.

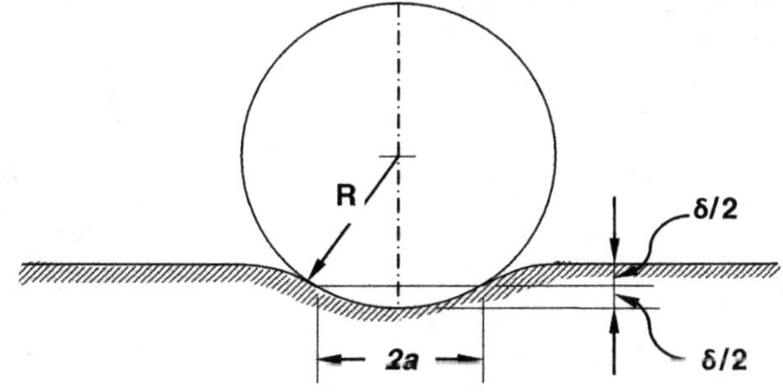

Fig. 3.10 Elastic contact of a smooth sphere and a flat surface

(*If we eliminate* p_0 *between the two equations, we get* $a^2 = R\delta$*, as before*).
But $p_0 = (3/2)p_m$, where p_m is the mean pressure.
Therefore

$$\delta = \frac{3\pi a p_m}{4E^*}$$

and

$$a = \frac{3\pi R p_m}{4E^*}$$

Eliminating a between the two equations, we get

$$\delta = \frac{9\pi^2 p_m^2}{16(E^*)^2} R$$

Basing his analysis on the von Mises maximum distortion energy criterion, Tabor (1951, p. 46) determined that, in this case, the onset of plastic deformation occurs when $p_m \approx 1.1\text{Y}$.

Substituting 1.1Y for p_m, we get the *maximum* elastic deformation:

$$\delta_e = 6.717 \frac{Y^2 R}{(E^*)^2}$$

When the deformation is fully plastic (see Fig. 3.11 and also, Mikic, 1974),

$$a^2 = 2R\delta$$

Also, according to Tabor (1951, p. 51), $p_m = 3Y$. Therefore,

$$\delta_p = \frac{a^2}{2R} = \frac{9\pi^2}{16(E^*)^2} \times \frac{9Y^2}{2R}$$

This simplifies to

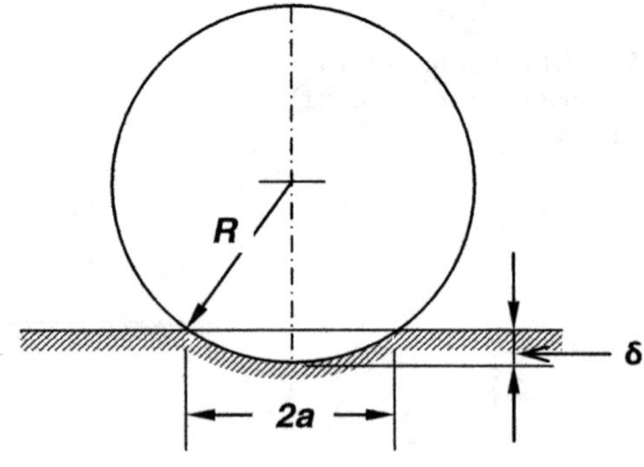

Fig. 3.11 Plastic contact of a smooth sphere and a flat surface

$$\delta_p = 24.98\frac{Y^2 R}{(E^*)^2}$$

for fully plastic deformation.

For deflections $\delta_e < \delta < \delta_p$, the radius of contact a and the mean pressure p_m may be obtained by interpolation:

$$a^2 = \left[\frac{\delta(2\delta_p - \delta_e) - \delta_e\delta_p}{\delta_p - \delta_e}\right]R$$

$$p_m = \left(\frac{\delta - \delta_e}{\delta_p - \delta_e}\right)(H - 1.1Y) + 1.1Y$$

The procedure for estimating the contact parameters and the solid spot conductance may be understood with reference to the flow chart (Fig. 3.12) shows.

Referring to the flow chart, the input parameters include the material properties, the applied load and the roughness test data. The roughness test data contains the distribution of the number of peaks against the peak height and the curvature against peak height and the slope for each surface. Note that the radius of curvature is the reciprocal of curvature. In the next step, the surface data are combined into one equivalent surface. The discretization process divides the surface peak height distribution into a large number, 400 for example, of sections. The maximum elastic and the minimum plastic deflection are calculated for each section (they are dependent on the asperity tip radius). An initial mean separation is set to start the iterative procedure. The deflection is compared with δ_e and δ_p values for this section to determine the deformation mode for this section. The number of peaks crossed, the contact radius and the load sustained are calculated based on this deformation mode and the deflection.

When the deformation mode is plastic, the microhardness value used is according to the size of the contact radius, as shown in Fig. 3.13. The sustained load is compared with the applied load. If they do not agree, the assumed deflection is changed and the procedure is repeated until the two values agree within a set limit. The thermal conductance is calculated, using Eq. (3.4), from the estimated radii of contacts, their number and the effective thermal conductivity. The calculations are then repeated for another load and another assumed deflection

The procedure showed promising results when tested against limited experimental data on thermal contact conductance by Li et al. who used an A/D converter as the interface between the surface analyser and a personal computer for digitising the measurements and for implementing the procedure described. With the availability of surface measuring equipment and hardness meters with built-in software, this procedure is now more readily able to be applied for the prediction of contact parameters and thermal conductance.

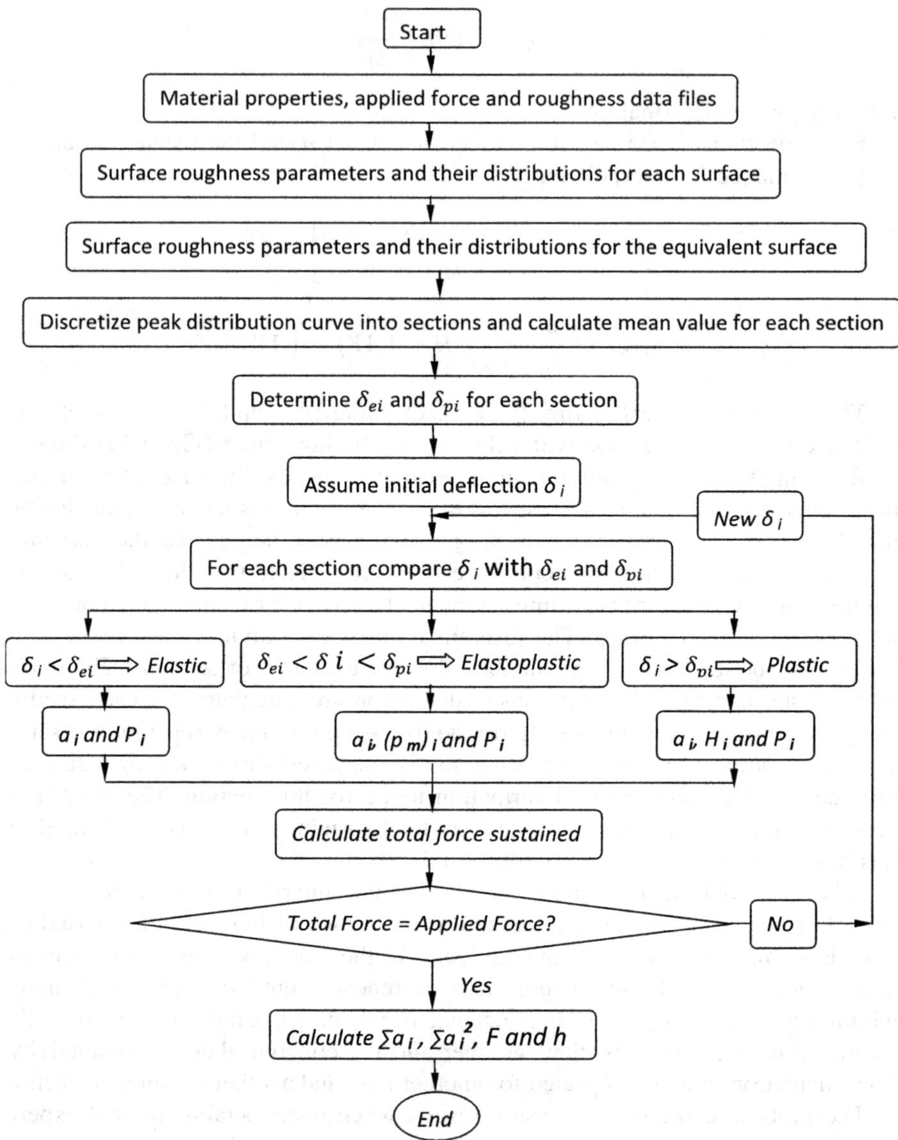

Fig. 3.12 Flow chart for determining contact parameters and solid spot conductance

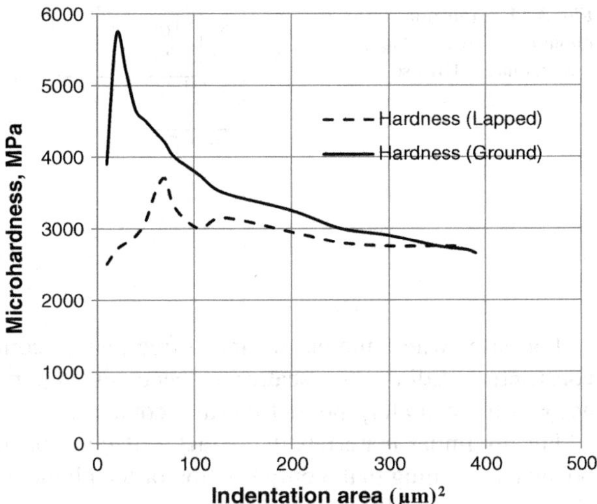

Fig. 3.13 Variation of hardness with indentation size (Stainless steel)

3.9 Thermal Boundary Resistance

With reference to heat flow across joints, "thermal boundary resistance" has been a topic of much speculation and research in recent years. Because of its importance to nanoscale systems, we present here a very brief introduction to this aspect of interface heat transfer.

In metals, conductivity is primarily due to free electrons. Thermal conductivity of metals is approximately proportional to the product of the absolute temperature and the electrical conductivity (Wiedemann–Franz law). In pure metals, the electrical conductivity decreases with increasing temperature and, therefore, the thermal conductivity stays approximately constant. This observation applies only to ordinary, and not ultra low, temperatures. In alloys, the change in electrical conductivity is usually smaller and thus thermal conductivity increases with temperature, often proportional to temperature.

On the other hand, thermal conductivity in non-metals is mainly due to lattice vibrations or phonons. Except for high quality crystals at low temperatures, the phonon mean free path is not reduced significantly at higher temperatures. Thus the thermal conductivity of non-metals is approximately constant at low temperatures. In dielectric materials there are few free electrons and heat conduction occurs primarily due to phonons.

Thermal boundary resistance (also known as Interfacial thermal resistance, or Kapitza resistance), differs from contact resistance, as it exists even at atomically perfect interfaces. Due to the differences in electronic and vibrational properties in different materials, when an energy carrier (phonon or electron, depending on the material) attempts to traverse the interface, it will scatter at the interface. The probability of transmission after scattering will depend on the available energy states on side 1 and side 2 of the interface.

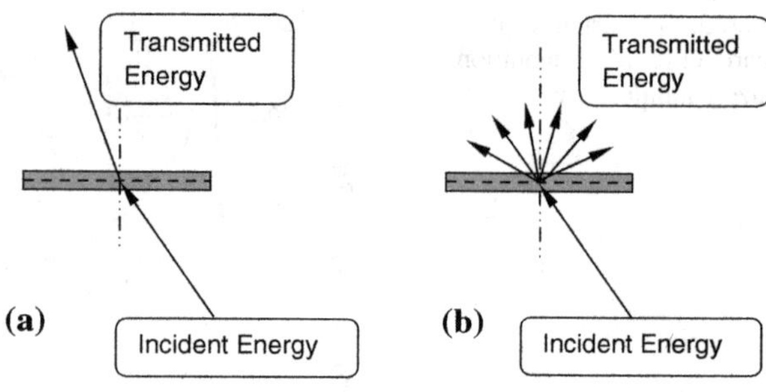

Fig. 3.14 Transmission of phonons at an Interface. **a** Specular. **b** Diffuse

For cases where the carrier mean free path is comparable to or larger than the constriction radius, this scattering becomes important and must be taken into account to accurately predict thermal contact resistance.

For dissimilar materials in contact, reflection of phonons will take place at the boundary meaning that a only fraction of the phonons will be transmitted from one side to the other (Fig. 3.14).

To calculate the boundary resistance in this case there are two widely used models: the acoustic mismatch model (AMM) and the diffuse mismatch model (DMM) . The AMM assumes a geometrically perfect interface and phonon transport across it is entirely elastic, treating phonons as waves in a continuum, that is, the transmission is specular. On the other hand, DMM assumes that all phonons are diffuse which is accurate for interfaces with characteristic roughness at elevated temperatures. (See Fig. 3.15).

As just stated, thermal boundary resistance is due to scattering of the carriers at an interface. The type of carrier scattered will depend on the materials governing the interfaces. For example, at a metal-to-metal interface, *electron* scattering effects will dominate thermal boundary resistance. If only one of the materials is a metal, then the resistance due to electron-phonon coupling should also be considered.

3.9.1 A Simple Model for Total Interface Resistance

Following an analysis similar to that for electron conductivity, the phonon thermal conductivity may be written as;

$$\kappa = \frac{1}{3} C_v^{ph} v L^{ph}$$

where C_v^{ph} is the specific heat capacity of phonons, L^{ph} is the mean free path phonons and v is the phonon velocity.

The following discussion is based mainly on the paper by Prasher and Phelan (2006).

In the completely ballistic limit, the constriction resistance, R_c, can be calculated using the method to calculate the flow rate of gas molecules through an orifice in the free molecular flow regime. Free molecular flow of gases is

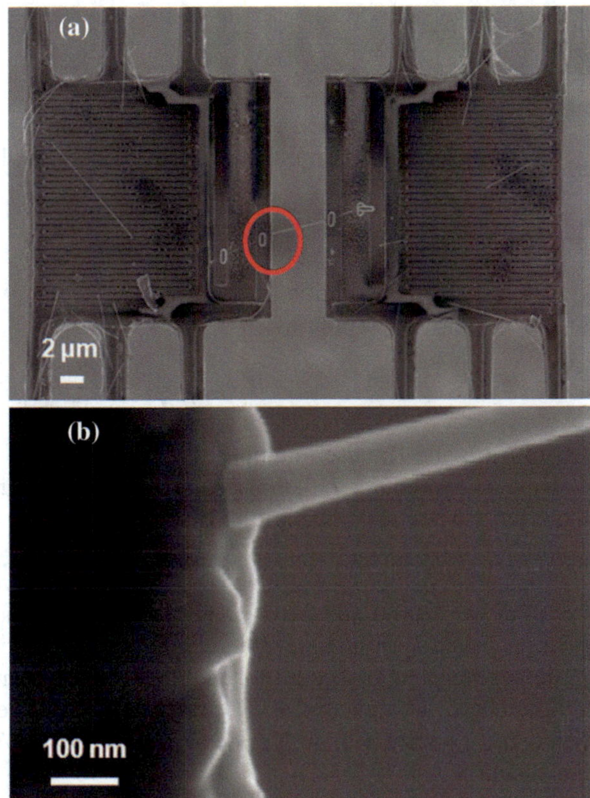

Fig. 3.15 **a** Low magnification SEM image of InAs NW 1 bridging two SiNx membranes. **b** High resolution SEM of the left contact between the NW and the membrane (Zhou et al. 2011; reproduced by the permission from American Institute of Physics)

analogous to the ballistic transport of phonons because both correspond to the case where the mean free path is much larger than the characteristic dimension of the system under consideration. In calculating R_c in the ballistic regime, phonons are considered as the heat carriers; however, the equations are equally valid for electrons.

If both sides of the interface are made of the same material, then the heat flux is:

$$q = \frac{vC_v^{ph}}{4}(T_1 - T_2)$$

Therefore, the ballistic constriction resistance, *based on unit area*, is;

$$R''_{cb} = \frac{4}{vC_v^{ph}}$$

But the thermal conductivity of the phonons is

$$\kappa = \frac{1}{3}C_v^{ph}vL^{ph}$$

Hence

$$R''_{cb} = \frac{4L^{ph}}{3\kappa}$$

For a constriction of radius a, (and area πa^2) the ballistic resistance is

$$R_{cb} = \frac{4L^{ph}}{3\pi a^2 \kappa}$$

Adding the constriction resistance [Eq. (3.1), with F neglected], the total resistance may be written as:

$$R_c = \frac{1}{2\kappa a}\left[1 + \frac{8}{3\pi} Kn\right]$$

where Kn is the Knudsen number = L^{ph}/a.

Note that when $a \gg L^{ph}$, the second term becomes insignificant and for $L^{ph} \gg a$, the first term becomes negligible. Both terms need to be considered when L^{ph} and a are of similar magnitude.

The above model for ballistic resistance assumes very strong bond at the interface. Prasher (2009) noted that nano particles are typically in contact with another surface through weak van der Waal (vdW) forces and presented an analytical model of the resistance that account for the strength of the interfacial bonding.

Zhou et al. (2011) used the DMM to calculate the phonon transmission coefficient ($\alpha_{1\to2}$) from Silicon Nitride (SiNx) to Indium Arsenide (InAs) according to the equation:

$$\alpha_{1\to2} = \frac{C_2 v_2}{C_1 v_1 + C_2 v_2}$$

where C and v are the specific heat and phonon group velocity and the subscripts denote the two materials in contact (Table 3.5).

They obtained a value of about 0.6 (actually 0.679) for $\alpha_{1\to2}$

The interface resistance was then calculated according to

$$R_i = \frac{4}{(\alpha_{1\to2} C_1 v_1 A)}$$

in which A is the contact area between the InAs nanowire (diameter 70 nm based on the high resolution SEM image, see Fig. 3.15) and the SiNx substrate. The R_i predicted by the above equation is 4.6×10^6 and 9.1×10^5 K/W for the estimated contact areas of about 130×10^{-18} and $650 \times 10^{-18}\text{m}^2$, respectively. These R_i values are two to three orders of magnitude smaller than the measured R_c of about 6×10^8 K/W.

Table 3.5 Physical properties of InAs and SiNx

Property	InAs	SiNx
Heat capacity, C [J/(m^{3}K)]	1.8×10^6	1.9×10^6
Sonic velocity, v [m/s]	5.5×10^3	11×10^3

These results confirmed that the DMM model is applicable only to the case of strong interface bonding, so-called welded contacts, that require the continuity of stress and displacement at the interface. The adhesion energy for vdW contacts between carbon nanotubes and a glass substrate has been determined to be less than 100 mJ/m^2, two orders of magnitude lower than those for welded contacts. Consequently, the transmission coefficient for the vdW contact could be about 100 times smaller than that calculated using DMM for a welded contact.

At temperatures well above 1 K, the AMM could describe the experimental results up to 40 K. However, at high temperatures the DMM is more appropriate than the AMM to describe the interface thermal conductivity. At intermediate temperatures, where both specular and diffuse transmission are likely, neither the AMM nor the DMM can describe accurately the interface thermal conductance. Kazan (2011), assuming a Gaussian probability density for the height and a two-dimensional tangential autocorrelation, presented a model for the interface thermal conductance that interpolates between the AMM and the DMM.

Thermal boundary resistance can significantly affect the properties of nano-materials when bundled together or in contact with other surfaces. It will be seen in a later chapter that although carbon nano tubes have high thermal conductivity, the thermal boundary resistance reduces their effective conductivity when these tubes are used in an array as an interstitial material.

References

Antonetti VW, Whittle TD, Simons RE (1993) An approximate thermal contact conductance correlation. ASME, Electronic Packaging 115:131–134

Bahrami M, Culham JR, Yovanovich MM (2004) Modeling thermal contact Resistance: a scale analysis approach. Trans ASME, J Heat Transfer 126:896–905

Bush AW, Gibson RD (1979) A theoretical investigation of thermal contact conductance. Appl Energy 5:11–22

Clausing AM, Chao BT (1965) Thermal contact resistance in a vacuum environment. Trans ASME, J Heat Transfer 87:243–251

Das AK, Sadhal SS (1999) Thermal constriction resistance between two solids for random distribution of contacts. Heat Mass Transf 35:101–111

Field JS, Swain MV (1993) A simple predictive model for spherical indentation. J Mater Res 8:297–305

Fletcher LS, Gyorog DA (1971) Prediction of thermal contact conductance between similar metal surfaces. Prog Astro Aero 24:273–288

Greenwood JA (1967) The area of contact between rough surfaces and flats. Trans ASME, J Lub Technol 89:81–91

Greenwood JA, Tripp JH (1970) The contact of two nominally flat rough surfaces. Proc I Mech Eng 185:625–633

Hunter AJ (1972) On the distribution of asperity heights of surfaces. J Phys D: Appl Phys 5:319–325

Ju Y, Farris TN (1996) Spectral analysis of two—dimensional contact problems. Trans ASME, J Heat Transfer 118:320–328

Kazan, M (2011) Interpolation Between the Acoustic Mismatch Model and the Diffuse Mismatch Model for the Interface Thermal Conductance: Application to InN/GaN Superlattice, Trans ASME, J Heat Trans, 133:112401 (7 Pages)

Kimura Y (1970) Estimation of the number and the mean area of real contact points on the basis of surface profiles. Wear 15:47–55

Lambert MA, Mirmira SR, Fletcher LS (2006) Design graphs for thermal contact conductance of similar and dissimilar light alloys. J Thermophys Heat Transfer 20:809–816

Laming LC (1961) Thermal conductance of machined metal contacts. In: Proceedings of the International Conference Devel Heat Transfer. American Society of Mechanical Engineers, New York, pp. 65–76

Laraqi N (2003) Thermal resistance for random contacts on the surface of a semi-infinite heat flux tube. Trans ASME J Heat Transfer 125:532–535

Li YZ, Madhusudana CV, Leonardi E (1998) Experimental investigation of thermal contact conductance: variation of surface microhardness and roughness. Int J Thermophysics 19:1691–1704

Li YZ, Madhusudana CV, Leonardi E (2000) On the contact of two isotropic rough surfaces: effect of microhardness variation. Mech Eng Trans, IE Aust, ME24:1–7

Mal'kov VA (1970) Thermal contact resistance of machined metal surfaces in a vacuum environment. Heat Transfer—Soviet Res 2(4):24–33

Mikic BB (1974) Thermal contact conductance: theoretical considerations. Int J Heat Mass Transfer 17:205–214

Mikic BB, Rohsenow WM (1966) Thermal Contact Resistance. Mech Eng Dept Report No. DSR 74542–41, MIT, Cambridge

Popov VM (1976) Concerning the problem of investigating thermal contact resistance. Power Eng (NY) 14:158–163

Prasher R (2009) Acoustic mismatch model for thermal contact resistance of van der Waals contacts. Appl Phys Lett 94:041905 (3 pages)

Prasher RS, Phelan PE (2006) Microscopic and macroscopic thermal contact resistances of pressed mechanical contacts, J Applied Physics 100:063538 (8 pages)

Roca RT, Mikic BB (1971) Thermal contact resistance in a non-ideal joint. Mech Eng Dept Report No DSR 71821-77, MIT, Cambridge

Sneddon IN (1965) The relation between Load and Penetration in the axisymmetric boussinesq problem for a punch of arbitrary profile. Int J Eng Sci 3:47–57

Tabor D (1951) The hardness of metals. Oxford University Press, London, pp 46–51

Tabor D (1975) A simplfied account of surface topography and the contact between solids. Wear 32(2):269–271

Thomas TR, Sayles RS (1975) Random process analysis of the effect of waviness on thermal contact resistance. Prog Astro Aero 39:3–20

Tien CL (1968) A correlation for thermal contact conductance of nominally-flat surfaces in vacuum. In: Proceedings of 7th Conference on Thermal Conductivity, U.S. National Bureau of Standards, Gaithersburg

Timoshenko SP, Goodier JN (1970) Theory of elasticity, 3d edn. McGraw-Hill, New York

Tsukizoe T, Hisakado T (1965) On the mechanism of contact between metal surfaces—the penetrating depth and the average clearance. Trans ASME, J Basic Eng 87:666–674

Tsukizoe T, Hisakado T (1968) On the mechanism of contact between metal surfaces. Part 2—the real area and the number of contact points. Trans ASME, J Lub Technol 90:81–88

Whitehouse DJ, Archard JF (1970) The properties of random surfaces of significance in their contact. Proc R Soc (London) Ser A 316:97–121

Yovanovich MM (1969) Overall constriction resistance between contacting rough, wavy surfaces. Int J Heat Mass Transfer 12:1517–1520

Zhan L (2001) Solid mechanics for engineers. Palgrave, Basingstoke (UK), p 320

Zhou F, Persson A, Samuelson L, Linke H, Shi L (2011) Thermal resistance of a nanoscale point contact to an indium arsenide nanowire. Appl Phys Lett 99:063110 (3 pages)

Chapter 4
Gap Conductance at the Interface

At low contact pressures (of the order of 10^{-4} H or less), it can be shown that the heat transfer across a joint occurs mainly through the gas gap (Madhusudana 1993). Boeschoten and van der Held (1957) also observed that the heat transfer was predominantly through the gas gap for "low [up to several kg/(sq cm)]" contact pressure. Lang (1962) pointed out that convection heat transfer is usually negligible for gap widths of up to about 6 mm (corresponding to a Grashof number of about 2000 for air at atmospheric pressure and a temperature of 300 K). Since the mean separation between contacting solid surfaces is some three orders of magnitude smaller than this dimension, it is clear that convection cannot be the mode of heat transfer across the gap. We conclude that the heat transfer across the gas-filled voids, interspersed between the actual contact spots, is principally by conduction, as alrcady notcd.

4.1 Factors Affecting Gas Gap Conductance

The following is a brief discussion of the factors affecting the heat transfer across the thin gas gap by conduction. A review summarising the state of knowledge on gas gap conductance, to 1980, was published by Madhusudana and Fletcher (1981).

If the gas layer can be considered as a continuum, then Fourier's law of heat conduction applies and the heat transfer coefficient, h_g, for the gas gap may simply be written

$$h_g = k_g/\delta \tag{4.1}$$

where

k_g = thermal conductivity of the gas

δ = mean thickness of the gas gap.

The effective thickness of the gap would be of the same order of magnitude as the surface roughness heights. For most practical surfaces in contact, the surface roughness heights have a range between 0.1 and 10 μm. The mean gap thickness

C. V. Madhusudana, *Thermal Contact Conductance*,
Mechanical Engineering Series, DOI: 10.1007/978-3-319-01276-6_4,

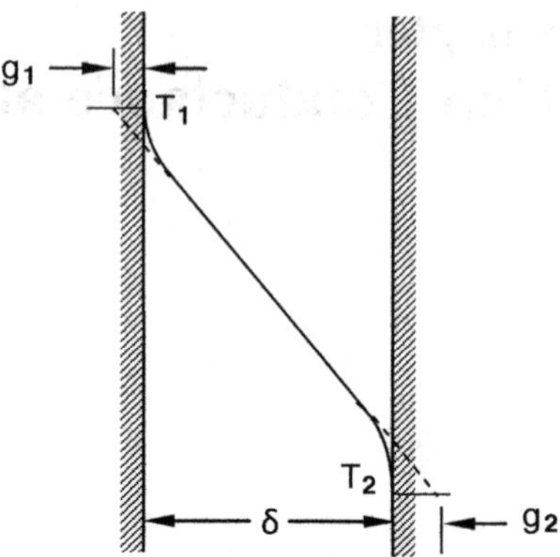

Fig. 4.1 The temperature jump distance

of smooth surfaces, therefore, would be similar in magnitude to the mean free path, λ, of the gas molecules at atmospheric pressure. Under these circumstances, the phenomenon of "temperature jump" becomes important. Since λ increases as the gas pressure is reduced, this effect becomes important for rough surfaces also if the pressure is sub-atmospheric.

The temperature jump accounts for the inefficiency in energy transfer between the gas molecules and the solid surfaces during a single collision (Smoluchowski effect). Thus when heat is conducted across two parallel plates separated by a distance similar to λ, the temperature distribution within the layer would be as shown in Fig. 4.1. It is clear from the figure that the effect of temperature jump is to increase the length of the heat transfer path by an amount that can be called the *temperature jump distance*. Kennard (1938) gives the following equation for the temperature jump distance:

$$g = \left(\frac{2-\alpha}{\alpha}\right)\left(\frac{2}{\gamma+1}\right)\left(\frac{k_g}{\mu C_v}\right)\lambda \tag{4.2}$$

In Eq. (4.2),
α = thermal accommodation coefficient
γ = ratio of specific heats for the gas, C_p/C_v
k_g = thermal conductivity of the gas
μ = viscosity of the gas
C_v = specific heat at constant volume
λ = mean free path of gas molcules
Since Prandtl Number, $\Pr = (\mu C_p/k_g)$, Eq. (4.2) is sometimes written as:

$$g = \left(\frac{2-\alpha}{\alpha}\right)\left(\frac{2\gamma}{\gamma+1}\right)\left(\frac{1}{Pr}\right)\lambda \tag{4.2a}$$

Equation (4.2) applies to a single gas. Vickerman and Harris (1975) determined the temperature jump distance, g_m for a *mixture of gases* to be

Table 4.1 Thermophysical properties of selected gases

Gas	k_g [W/(mK)]	γ	μ [10^{-6} kg/(ms)]	C_v [J/(kgK)]	λ (10^{-6} m)
Hydrogen	0.180	1.41	8.9	10120	0.118
Helium	0.149	1.66	19.8	3150	0.186
Neon	0.048	1.64	31.6	635	0.132
Nitrogen	0.026	1.40	17.8	741	0.063
Oxygen	0.0267	1.40	20.7	657	0.068
Argon	0.0167	1.67	22.4	310	0.067
Carbon dioxide	0.0167	1.30	14.9	648	0.042
Air	0.0262	1.40	18.5	718	0.064

$$g_m = \frac{\sum\left(\frac{x_i g_i}{M_i^{0.5}}\right)}{\sum\left(\frac{x_i}{M_i^{0.5}}\right)} \tag{4.3}$$

where

x_i = mass fraction of constituent gas i

M_i = molecular mass of constituent gas i

g_i = temperature jump distance of constituent gas i.

Table 4.1 lists the above properties for several gases. The values refer to atmospheric and temperature of approximately 101 kPa and 300 K, respectively. Since the accommodation coefficient is a composite property that depends on the nature of the gas as well as the solid surfaces with which the gas is in contact, it is discussed separately in the next section.

Equation (4.1) may now be modified, as follows, to take the temperature jump distance into account:

$$h_g = k_g/(\delta + g_1 + g_2) \tag{4.4}$$

Thus the problem of determining the gas gap thermal conductance reduces to one of determining the mean gap thickness and the temperature jump distance.

When dealing with gas gap conductance, a non-dimensional parameter called the Knudsen number, N_{Kn} as defined below, is useful:

$$N_{Kn} = \lambda/\delta \tag{4.5}$$

For convenience, we can then identify three regimes of gas gap conduction:

1. Continuum: $N_{Kn} \ll 1$. Fourier's law of heat conduction may be applied.
2. Temperature Jump: $0.01 < N_{Kn} < 10$; Eq. (4.4) is applicable.
3. Free molecular conduction: $N_{Kn} > 10$; in this case, the mean physical gap thickness would be much smaller than the temperature jump distances and Eq. (4.4) may be approximated as:

$$h_g \cong k_g/(g_1 + g_2) \tag{4.6}$$

In other words, the heat transfer rate would be independent of the physical distance between the two surfaces.

4.2 The Accommodation Coefficient

It is clear from Eq. (4.2) that the accommodation coefficient plays an important role in controlling the temperature jump distance and the gas gap conductance. Accommodation coefficient stands for the fractional extent to which the molecules, that fall on the surface and are reflected or re-emitted from it, have their energy adjusted or accommodated toward what the energy would have been if the returning molecules were issuing as a stream of gas at the temperature of the wall. In other words, accommodation coefficient characterises the extent of gas-surface energy exchange. It is, therefore, defined as:

$$\alpha = \frac{T_f - T_g}{T_s - T_g} \tag{4.7}$$

in which

T_s = temperature of the surface
T_g = temperature of the incident gas
T_f = effective temperature of the scattered gas.

It is evident that the accommodation coefficient, at a given temperature, must depend upon the nature of both the gas and the (solid) surface. There exists a large number of works dealing with the experimental and theoretical determinations of the accommodation coefficient for various "gases" in contact with specific solid materials (see, for example, Wiedmann and Trumpler 1946; Wachman 1962; Semyonov et al. 1984).

The following discussion should provide some insights to the nature and behaviour of the accommodation coefficient (see Dharmadurai 1983). Based on phonon energy transmission theory, he derived the following expression for the accommodation coefficient of a *monatomic* gas:

$$\alpha \cong \frac{2n\left(\frac{A_e}{A}\right)\left(\frac{M_g}{M_s}\right)}{1 + n\left(\frac{A_e}{A}\right)\left(\frac{M_g}{M_s}\right)} \tag{4.8}$$

at ambient temperatures in contact with a *clean* solid surface.

In Eq. (4.8)

n = atomicity (valency) of the solid
A_e = effective surface area available for gas/surface energy interchange
A = macroscopic interfacial area
M_g = molecular mass of gas
M_s = molecular mass of solid.

For smooth surfaces, in general, $\left(\frac{A_e}{A}\right) = 1$; for rough surfaces, the ratio can be significantly larger than 1. (Even for smooth surfaces, $\left(\frac{A_e}{A}\right)$ can be larger than 1, if the atomic diameter of the gas is much smaller than that of the solid) It can be seen that, for clean solid surfaces in contact with a light monatomic gas for which

$\left(\frac{M_g}{M_s}\right) \ll 1$, Eq. (4.8) yields a value of $2\left(\frac{M_g}{M_s}\right)$ for the accommodation coefficient. This is in accordance with the rigorous gas-scattering theories.

For *contaminated* surfaces, the above equation is modified as:

$$\alpha \cong \frac{2\left(\frac{M_g}{A}\right)\left\{fC_a n_a\left(\frac{A_{ea}}{M_a}\right) + (1-f)\left(\frac{n_s A_{es}}{M_s}\right)\right\}}{1 + \left(\frac{M_g}{A}\right)\left\{fC_a n_a\left(\frac{A_{ea}}{M_a}\right) + (1-f)\left(\frac{n_s A_{es}}{M_s}\right)\right\}} \quad (4.9)$$

In the above expression, C_a is a constant characterising the condensed phase of the adsorbed molecules; f is the fraction of occupied surface sites and the subscripts a and s refer to the absorbed and the surface molecules respectively. If the condensed phase is solid-like, then $C_a = 1$, while if it is fluid-like, then $C_a \approx 0.5$. For clean surfaces $f = 0$.As the temperature of an initially totally covered surface ($f = 1$) rises, desorption increases leading to a gradual decrease in f. It follows that for systems with $(M_s)/A_{es} > M_a/(C_a A_{ea})$, the accommodation coefficient should decrease as the temperature increases.

Song and Yovanovich (1987) developed a correlation for the accommodation coefficient for "engineering" surfaces (that is, surfaces with adsorbed layers of gases and oxides). This correlation was based on the experimental results of several previous investigators, for monatomic gases. The resulting relation was extended by the use of a "monatomic equivalent molecular mass" to apply for diatomic/polyatomic gases. The final correlation is given below:

$$\alpha = \exp(C_0 T)\left[\left(\frac{M_g}{C_1 + M_g}\right) + \{1 - \exp(C_0 T)\}\frac{2.4\beta}{(1+\beta)^2}\right] \quad (4.10)$$

where

C_0 = dimensionless constant equal to –0.57

T = $(T_s - T_O)/T_O$; (T_0 = 273 K)

$M_g = M_g$ = for monatomic gases

1.4 M_g for diatomic/polyatomic gases

C_1 = 6.8, units of M_g (g/mole)

$\beta = M_g/M_s$

The agreement between the published experimental data for diatomic and polyatomic gases and the predictions according to the above correlation was generally within ±25 %.

Table 4.2 summarises the accommodation coefficients for single gases in contact with various metals, as determined experimentally by different investigators. Table 4.3 lists the accommodation coefficients for air determined by Wiedman and Trumpler (1946). Unless otherwise noted, the values in both tables pertain to ambient temperatures of about 300 K.

Table 4.2 Accommodation coefficients for single gases

	He	Ne	Ar	H_2	O_2	N_2	CO_2	Reference
Platinum	0.38	0.75	0.8	0.24	0.62	0.68	0.52	Smoluchowski (1898)
Platinum (bright)	0.44			0.32				Knudsen (1934)
Platinum (blackened)	0.91			0.72				Knudsen (1934)
Platinum (uncleaned)	0.446	0.816	1.01			0.975	1	Semyonov et al. (1984)
Platinum			0.644					Thomas and Brown(1950)
Tungsten (clean 1000 °C)				0.54				Blodgett and Langmuir (1932)
Tungsten (clean fresh)	.06–.07							Roberts (1932)
Tungsten (uncleaned)	0.393	0.796	1			0.975	1	Semyonov et al. (1984)
Tungsten (uncleaned)	0.493	0.848	0.974			0.991	1	Semyonov et al. (1984)
Uranium dioxide	0.55		0.75					Hall et al. (1990)
Nickel (uncleaned)	0.457	0.831	1.02			0.978	1.02	Semyonov et al. (1984)

Table 4.3 Accommodation coefficients for air

Solid surface	Accommodation coefficient
Flat black lacquer on bronze	0.881–0.894
Bronze, polished	0.91–0.94
Bronze, machined	0.89–0.93
Bronze, etched	0.93–0.95
Cast iron, polished	0.87–0.93
Cast iron, machined	0.87–0.88
Cast iron, etched	0.89–0.96
Aluminium, polished	0.87–0.95
Aluminium, machined	0.95–0.97
Aluminium	0.89–0.97

Note The apparatus employed by wiedmann and Trumpler for determining the accommodation coefficients for air required the calculation of heat transfer by radiation and rarefied gas conduction between two long, concentric cylinders. They noted that the emissivity of etched aluminium dropped from 0.833 to 0.753 during the tests. Hence it was suggested that the accomodation coefficient data for etched aluminium was not reliable

4.2.1 Effect of Temperature on Accommodation Coefficient

The experimental results of Ullman et al. (1974) for helium and xenon in contact with stainless steel and uranium dioxide surfaces indicated that, in all cases, the accommodation coefficient decreased with temperature. The temperature range of the solid surfaces during their tests was 500–1000 K and no attempt had been made to clean or polish either of the surfaces. Their data for helium is approximated by the correlation (Thomas and Loyalka 1982)

$$\alpha = 0.425 - 2.3 \times 10^{-4}\, T$$

This is in agreement with the conclusion reached earlier in this section, namely, that the accommodation coefficient for *unclean* surfaces should decrease as the temperature increases. On the other hand, the results reported by Kharitonov et al. (1973) indicate that, for helium and neon gases in contact with *pure* tungsten, the accommodation coefficient increases with the temperature for temperatures greater than 300 K. A comparison of the two tests of data further shows that the accommodation coefficients for helium in contact with an unclean surface is at least an order of magnitude greater than those obtained with clean surfaces.

Before we leave this section on accommodation coefficients, it is interesting to consider the experimental observation of Cohen et al. (1960) that the conductance between the fuel and the jacket in a nuclear reactor was independent of the gas composition. An explanation for this unexpected behaviour was offered by Kharitonov et al. (1973), based on the fact [see Eqs. (4.5) and (4.6)] that the accommodation coefficient depends on the molecular mass. Thus, for a gas such as helium with low molecular mass, the accommodation coefficient would also be small, as seen in Table 4.2. For such a situation, one can write

$$g \cong \frac{\lambda}{\alpha} \tag{4.11}$$

Furthermore, the mean free pathfor helium is large and, for small gap thicknesses and/or low gas pressures the assumption of free molecular conduction, Eq. (4.6), is valid so that

$$h_g \cong k_g/(2g) \cong k_g\alpha/(2\lambda) \tag{4.12}$$

In Eqs. (4.11) and (4.12), it is assumed that the accommodation coefficients and, therefore, the temperature jump distances are the same for the two surfaces with which the gas is in contact.

From Eq. (4.12) we see that it would be wrong to assume that the gap conductance would be proportional to the gas conductivity alone; gap conductance is also affected by the ratio *(α/λ)*, which is comparatively small for helium. Indeed it can be shown that for pure surfaces of heavy metals, the gap conductance with xenon would be greater than that with helium in contact with a similar pair of surfaces, although the conductivity of helium is about thirty times that of xenon.

4.2.2 Summary of Observations

The following general conclusions may be made in view of the above discussion of previous experimental and theoretical investigations on accommodation coefficient.

Accommodation coefficients usually range in values from 0.01 to 1.0, although values higher than 1 are possible.

In general, lighter monatomic gases have low accommodation coefficients.

Clean surfaces have lower accommodation coefficients compared with contaminated surfaces.

For unclean surfaces, accommodation coefficient decreases as the temperature increases.

The accommodation coefficient appears to be inversely proportional to the thermal conductivity of the gas

The accommodation coefficient is independent of the gas pressure as it is mainly dependent on the relative molecular masses of the gas and the solid.

4.3 Effect of Gas Pressure on Gap Conductance

From kinetic theory of gases, it can be shown that the thermal conductivity of a gas is given by (Hardee and Green 1968):

$$k_g = \frac{4f}{3d^2}\left(\frac{k^2 T}{\pi^2 M}\right)^{0.5} \tag{4.13}$$

where

f = number of degrees of freedom of the gas molecule
d = diameter of the gas molecule, m
T = absolute temperature of the gas, K
k = Boltzman constant, 1.381×10^{-23}, J/K
M = molecular mass of gas, kg.

It can be seen that the thermal conductivity of a gas is independent of its pressure. Therefore, Aaron (1963) considered that the conductance in the gas gap should be insensitive to decreases in gas pressure until a certain "threshold pressure" is reached when the mean free path of the gas molecules become comparable in magnitude to the average gap thickness, and Eq. (4.4) becomes applicable. For Aaron's experiments in air with a gap thickness of 9.6 μm, this threshold pressure was found to be equal to 21 mm Hg (2.79 kPa) at a temperature of 300 K. The decrease in gas conductance was not noticeable until this pressure was reached—further decreases in the gas pressure would result in a decrease in gap conductance since the mean free path and, consequently, the temperature jump distance will increase. A similar conclusion can be drawn from the results of

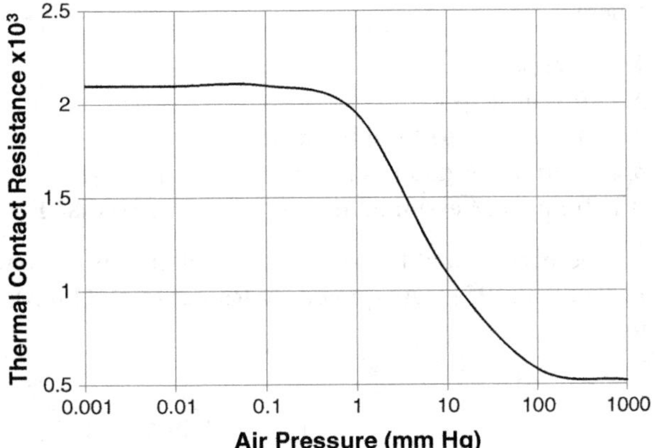

Fig. 4.2 Effect of air pressure on TCR (after Shlykov and Ganin, 1964)

Shlykov and Ganin's (1964) tests on stainless steel contacts in air (see Fig. 4.2). It can be seen that, for a given contact pressure (in this case 20 kg/ $cm^2 \approx$ 19.6 MPa), there was noticeable increase in the resistance only when the air pressure dropped below 100 mm Hg ($\approx$13 kPa). In other words, the threshold pressure applicable for this situation was 13 kPa. The experimental results of Madhusudana (1975) on stainless steel/Nilo contacts in air confirm the existence of a threshold pressure of similar magnitude.

At the other end of the spectrum, Shlykov and Ganin's results show that, for a given contact pressure, there is very little increase in the total contact resistance for gas gap pressures below 1 mm Hg, approximately (Fig. 4.2). They therefore concluded that the gas gap conductance became practically zero at a pressure of 0.1 mm Hg. Hence it was recommended that tests in a vacuum of the order of 0.1 mm Hg was sufficient to determine the solid spot conductance. Cassidy and Mark (1969) experimentally measured TCR of stainless steel (type AISI 416) joints in air as the ambient pressure was decreased from one atmosphere down to 3×10^{-12} mm Hg. They confirmed that the assumption of zero thermal conductance was valid for gas pressures below 1 mm Hg, The more recent theoretical and experimental results of Nishino and Tori (Nishino and Torii 1994) also confirm that the thermal contact conductance is insensitive to air pressures below 10 Pa.

4.4 Correlations for Gas Gap Conductance

From Eq. (4.4), it would appear that, in order to estimate the gap conductance for a given gas, it is desirable to correlate the mean gap width δ with some physically measured surface parameters and the applicable temperature jump distances. Several correlations have been proposed of the form:

$$Y = f(X) \tag{4.14}$$

where

$Y = b_t/\delta_{eff}$
$X = b_t/(g_1 + g_2)$
b_t = total peak-to-trough surface roughness
δ_{eff} = effective gap thickness = $(\delta + g_1 + g_2)$
It is frequently assumed that $g_1 = g_2 = g$ so that $Y = b_t/(\delta + 2g)$ and $X = b_t/2g$.

The earliest and the simplest correlation is the one proposed by Cetinkale and Fishenden (1951). As quoted by Rapier et al. (1963), this correlation is equivalent to:

$$Y = \frac{1}{0.305 + (1/X)} \tag{4.15}$$

The experimental results of Cetinkale and Fishenden pertained only to large values of X and, in fact, only confirm that $Y \approx 1/0.305$ in this range.

The correlation proposed by Rapier et al. (1963), based on the experimental results of several investigators, and their own results for helium, neon and argon in contact with uranium dioxide—stainless steel surfaces is:

$$Y = \frac{0.6}{1 + (1/2X)} + 0.4ln(1 + 2X) \tag{4.16}$$

Correlations in Eqs. (4.15) and (4.16) do not take into account the effect variation of gap thickness with contact pressure. Dutkiewicz (1966) conducted a numerical analysis of surfaces taking into account the variation of contact area and, therefore, the gap thickness with contact load. He assumed that the asperity heights were normally distributed. His result, presented in tabular form, give values of the gap thickness variable $D^* = \sigma/(\delta + 2g)$ for arbitrary values of the temperature jump distance variable $C^* = \sigma/l2g$, and the ratio B of real to apparent area of contact. In the expressions for C* and D^*, σ is the standard deviation of the profile height distribution. His results also showed that $b_t \cong 6\sigma$ which indicates that $C^* = X/6$ and $D^* = Y/6$ according to the nomenclature used in the present section. Table 4.4 lists the values obtained by Dutkiewicz for the case when the standard deviations for both the surfaces are the same.

Table 4.4 Variation of gap thickness variable D* with contact area

C*	B				
	0	0.01	0.025	0.050	0.100
200	0.290	0.435	0.531	0.713	2.593
20	0.279	0.410	0.499	0.653	1.178
2	0.237	0.315	0.367	0.431	0.531
0.2	0.111	0.124	0.131	0.138	0.147
0.02	0.019	0.019	0.020	0.020	0.020

Another correlation, which takes the variation of gap thickness with load, is that due to Popov and Krasnoborod'ko (1975). This follows along the lines of Rapier et al. with X_1 and Y_1 defined as follows:

$$X_1 = \delta_{max}\left(\frac{1-\zeta}{g_1+g_2}\right); \quad Y_1 = \delta_{max}\left(\frac{1-\zeta}{\delta_{eff}}\right) \tag{4.17}$$

where ζ is the relative approach of the two surfaces due to mechanical loading; δ_{max} is defined as the maximum thickness of the interface layer. A careful study of their analysis indicates that δ_{max} could be taken to be the same as b_t. Hence we see that, for zero load, X_1 and Y_1 correspond to X and Y, respectively of this section. Popov and Krasnoborod'ko provided empirical correlations between Y_1 and X_1 for steel and Duralumin for different classes of surface finish.

Figure 4.3 below compares the different correlations for zero mechanical loading. Curve 3, representing the prediction of Popov and Krasnoborod'ko, is actually a composite graph—it shows the average values for four different classes of surface finish, It has been plotted from the expressions given by them for steel surfaces. From an examination of Fig. 4.3, it is clear that there is not a great deal of difference between the four predictions. It should be emphasized, however, that the correlations of Cetinkale and Fishenden, and Rapier et al., do not allow for the variation of gap thickness with the contact pressure. Of course, whatever be the correlation used, it is first of all necessary to establish the surface parameters and the temperature jump distances before proceeding to determine the effective gap thickness.

For conforming rough surfaces with a Gaussian height distribution, the mean separation between the surfaces can be calculated using the correlation presented by Antonetti (1992):

$$\delta = 1.53\sigma(P/H)^{-0.097} \tag{4.18}$$

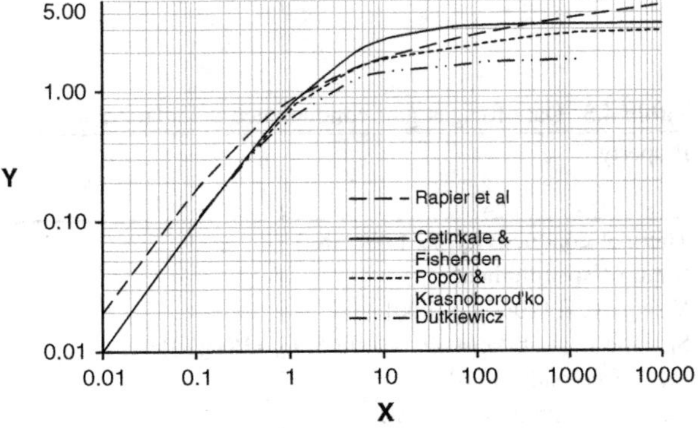

Fig. 4.3 Comparison of correlations for effective gas gap thickness

The gas gap conductance, at any mechanical load P, may be determined by using this equation in conjunction with Eq. (4.4).

Real surfaces have a finite maximum height whereas the normal distribution assumes an infinite maximum height. Therefore, Majumdar and Williamson (1990) suggested that the inverted Chi square (ICS), rather than Gaussian, distribution should be used for the surface heights. The use of ICS distribution results in higher values for gap conductance.

4.5 Numerical Example: Gas Gap Conductance

For the stainless steel/aluminium rough/rough surfaces (pair A of the numerical example of Chap. 3, the gap conductance is calculated below. Additional data is as shown in Table 4.5.

From Eq. (4.18), the mean physical gap at contact pressure P is

$$\begin{aligned}\delta &= 1.53\sigma(P/H)^{-0.097}\\ &= 1.53(1.77)(P/1400)^{-0.097}\end{aligned}$$

Hence $\delta = 5.468(P^{-0.097})$; δ will be in μm if P is expressed in MPa

From Eq. (4.2), the temperature jump distance is:

$$g = \left(\frac{2-\alpha}{\alpha}\right)\left(\frac{2}{\gamma+1}\right)\left(\frac{k_g}{\mu C_v}\right)\lambda$$

so that

$$g_{\text{air}} = \left(\frac{1.1}{0.9}\right)\left(\frac{2}{2.4}\right)\left(\frac{0.0262}{18.5\times10^{-6}\times718}\right)\times 0.064\times10^{-6} = 0.1286\ \mu m$$

and

Table 4.5 Data for gas gap conductance calculation

Property	Gas	
	Air	Helium
Thermal conductivity, k_g [W/(mK)]	0.0262	0.149
Ratio of specific heats, γ	1.40	1.66
Viscosity, μ [10^{-6} kg/(ms)]	18.5	19.8
Specific heat at constant voloume, C_v [J/(kgK)]	718	3150
Mean free path, λ (10^{-6} m)	0.064	0.186
Accommodation coefficient, α	0.90	0.45

Table 4.6 Gas gap and solid spot conductances compared

P, MPa	0.1	0.5	1	5	10	50	100
δ, μm	6.836	5.848	5.468	4.678	4.373	3.741	3.498
$h_{g,air}$, W/(m²K)	3693	4291	4576	5309	5658	6552	6977
$h_{g,helium}$, W/(m²K)	16460	18477	19391	21614	22612	25011	26076
h_s, W/(m²K)	626	2842	5453	24755	47494	215609	413652

$$g_{Helium} = \left(\frac{1.55}{0.45}\right)\left(\frac{2}{2.66}\right)\left(\frac{0.149}{19.8\times10^{-6}\times3150}\right)\times 0.186\times10^{-6} = 1.1508\ \mu\text{m}$$

The gap conductance is then given by

$$h_g = k_g/(\delta + 2g)$$

Hence Table 4.6 is generated. The solid spot conductance, h_s, from the numerical example in Chap. 3 is also included in the bottom row for the sake of comparison.

It is interesting to compare the above results with those obtained by using the correlation of Negus and Yovanovich (1988).

$$h_g = (k_g/\sigma)I_g \tag{4.19}$$

in which

$$I_g = \frac{f_g}{(\delta/\sigma) + (G/\sigma)}$$

where G is the sum of the temperature distances. So, we immediately note that Eq. (4.19) is the same as the basic Eq. (4.2), modified by the "correlation factor" f_g, Negus and Yovanovich recommended the following expressions for evaluating f_g:

$$f_g = 1.063 + 0.0471\left(4 - \frac{\delta}{\sigma}\right)^{1.68} ln\left(\frac{\sigma}{G}\right)^{0.84} ;for\quad 0.01 \le (G/\sigma) \le 1$$

and

$$f_g = 1 + 0.66(\sigma/G)^{0.8};\ for\ (G/\sigma) > 1$$

They also suggested using

$$\left(\frac{\delta}{\sigma}\right) = \sqrt{2}\,erfc^{-1}\left(\frac{2P}{H}\right)$$

for calculating the load-dependent separation between the surfaces, rather than Eq. (4.18).

In the above numerical example:

For air, $(G/\sigma)_{\text{air}} = 2 \times \frac{0.1286}{1.77} = 0.1453$ and for helium, $(G/\sigma)_{\text{Helium}} = 2 \times \frac{1.108}{1.77} = 1.252$

We compare the results at P = 5 MPa.

$$(\delta/\sigma) = \sqrt{2}\, erfc^{-1}(2 \times 5/1400) = \sqrt{2}\, erfc^{-1}(0.00714) = 2.701$$

For air:

$$f_{g,\text{air}} = 1.063 + 0.0471(4 - 2.701)^{1.68} ln\left(\frac{1}{0.1453}\right)^{0.84} = 1.181;$$

For helium:

$$f_{g,\text{Helium}} = 1 + 0.06(1/1.252)^{0.8} = 1.050$$

Thus we get, at P = 5 MPa,

$$h_{g,\text{air}} = \left(0.0262/1.77 \times 10^{-6}\right) \times \frac{1.181}{2.701 + 0.1453} = 6142\left[\frac{W}{m^2K}\right]$$

and

$$h_{g,\text{Helium}} = \left(\frac{0.149}{1.77} \times 10^{-6}\right) \times \frac{1.050}{2.701 + 1.252} = 22360\left[\frac{W}{m^2K}\right]$$

These compare with the figures of 5309 and 21614 for air and helium, respectively obtained by the first method. The differences 13.6 and 3.1 %, respectively are mainly due to the "correlation factor", f_g which has a value slightly greater than 1. In the original form of the gap conductance equation, Eq. (4.2), f_g is equal to 1.

The calculations emphasize that:

1. The gas gap conductance is the predominant mode of heat transfer at low to moderate contact pressures of up to 1 MPa. It could be significant at higher pressures too, depending on the gas/solid combination.
2. The variation in gas gap conductance with contact pressure is relatively small; for a thousand fold increase in contact pressure, the gap conductance increases by less than two fold.
3. The gas gap conductance is not directly proportional to the gas thermal conductivity.

Summaries of correlations for gap conductance may be found in Song et al. (1993) and Yovanovich and Marotta (2003).

4.6 Recent Research in Gap Conductance

Results of an extensive experimental programme on the gap conductance of gases and gas mixtures have been reported by Wahid and Madhusudana (2000). The gases used were helium, argon, nitrogen and carbon dioxide. Note: A separate, constant temperature mixing chamber was used to mix the gases in the required proportions.

Table 4.7 below lists the relevant details of the experiments. Table 4.8 lists the gas combinations together with the effective rms roughness.

Table 4.7 Experimental Details

Solids in contact	AISI 304 stainless steel
Contact pressure	0.433 MPa
Vacuum level in the chamber (to determine solid spot conductance)	3×10^{-2} mbar
Gas pressure	0.12 MPa (slightly above atmospheric)
Experimental accuracy	3 % in vacuum; 5–12 % otherswise

Table 4.8 Gas combinations and roughness of stainless steel surfaces

Interfacial gas or gas mixture							Rq (μm)
He	He:Ar	He:Ar	He:Ar	Ar	N_2	CO_2	
100	75:25	50:50	25:75	100			5.4
100	75:25	50:50	25:75	100	100	100	14.3
100	75:25	50:50	25:75	100	100	100	21.2

The gap conductance was measured as the difference between the total conductance (as measured in the gaseous atmosphere) and the solid spot conductance (as measured in vacuum). Then, in each case, the effective gap width was obtained by dividing the corresponding thermal conductivity by the conductance. The mean separation was then deduced by subtracting the temperature

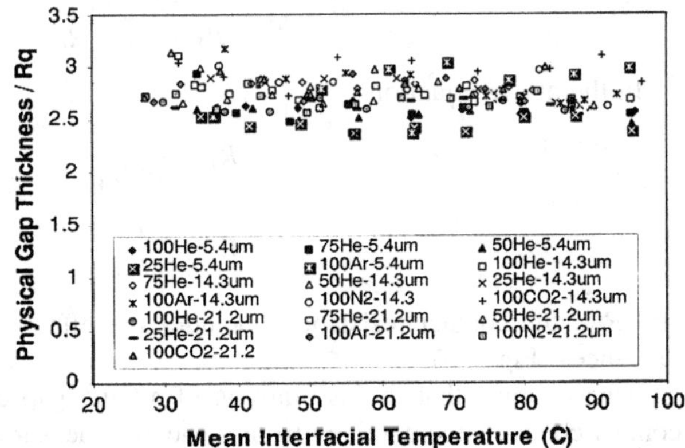

Fig. 4.4 Non dimensional mean gap thickness vs temperature (Wahid and Madhusudana, Reprinted by permission from Pergamon)

jump distances from the effective gap width. Finally, the mean separation was non-dimensionalised by dividing it by the surface roughness. The results are shown plotted in Fig. 4.4.

The following simple correlation for the mean physical gap is evident from the figure:

$$\delta \approx 2.7 R_q \tag{4.20}$$

It is noted that 85 % of the experimental results (173 data points, comprising three different surface roughnesses and seven different gases and gas mixtures) fall within ±4 % of this correlation. The correlation may be compared with those of Tsukizoe and Hisakado (1965) and Popov and Krasnoborod'ko (1975) both of whom suggested a value of 3.0 for the constant in Eq. (4.20). It may be stressed that the temperature jump distances, if significant, need to be added to δ to get the *effective* gap thickness.

A different theoretical approach for the determination of the contact conductance (including the solid spot and the gap conductances) was presented by Shaikh, Beall and Razani (Shaikh et al. 2001) who modelled the contact resistance between two dissimilar materials of finite thickness as two cylinders in contact over an area at their centres. The noncontact gap between the two cylinders was assumed to be filled with a thermally conducting fluid. The lateral surfaces were insulated, while the top and bottom surfaces were kept at constant temperatures. Heat diffusion equations in the cylinders were transformed to two integral equations for the heat flux through the contact and noncontact areas with the interstitial fluid conductance as a parameter. The integral equations were solved numerically. No assumption was made regarding on the heat flux distribution on the contact and noncontact surfaces.

The total resistance to heat flow was assumed to be the sum of the material resistances of the two cylinders, and the contact resistance due to the imperfect contact. (The gas or material present between the non-contacting surfaces will help to reduce the contact resistance). This way, the contact resistance was obtained by subtracting the effect of the material resistances of the two cylinders from the ratio of the temperature difference of the reservoirs to the total rate of heat transfer.

$$Q = \frac{T_2 - T_1}{R_1 + R_2 + R_{tot}}$$

In the above equation:

$$R_1 = \frac{c_1}{\pi b^2 k_1}$$

$$R_2 = \frac{c_2}{\pi b^2 k_2}$$

and R_{tot}, the contact resistance, includes the effects of solid spot and gap conductances (Fig. 4.5).

It was shown that the assumption of adding gap conductance to the value of contact conductance obtained from a vacuum measurement was reasonable for all

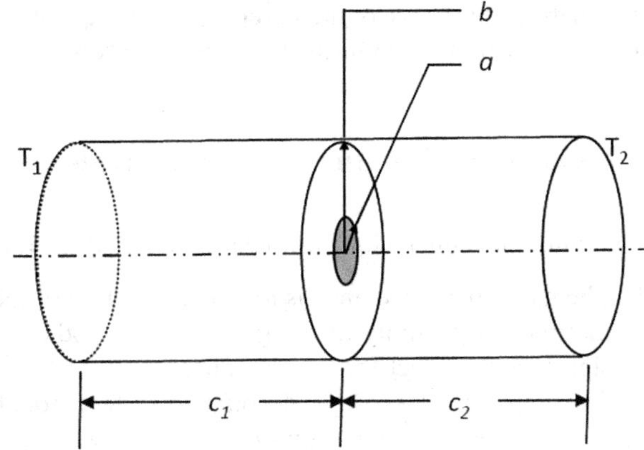

Fig. 4.5 Contact resistance model used by Shaikh et al. (2001)

cases of contacting materials including the finite geometries. It was also shown that the usual assumption of constant temperature on contacting surface was also valid, including the cases of interstitial fluid in the gap.

Misra and Nagaraju (2010) reported results of careful experiments on gold plated copper/copper and brass/brass contacts in vacuum, nitrogen and argon environments. The type of copper used was OFHC (oxygen-free high conductivity) and the thickness of coating ranged from 0.1 μm to 0.5 μm.

They observed that the gap conductance with coating was always lower than that without coating This was explained on the basis of the reduced accommodation coefficient when the material of the solid surface has a higher molecular mass than the bare surface. Misra and Nagaraju used the correlation due to Negus and Yovanovich, Eq. (4.10), to note that

For nitrogen in contact with copper, $\beta = 28/64 = 0.4375$ and, $2.4\beta/(1 + \beta)^2 = 0.5081$.

For nitrogen in contact with gold, $\beta = 28/197 = 0.1421$ and, $2.4\beta/(1 + \beta)^2 = 0.2615$.

Therefore the accommodation coefficient, and hence the gap conductance, for copper with gold coating will be smaller than the one without coating. Similar conclusions can be drawn when the environment is argon, The calculations, however, were not extended to evaluate the accommodation coefficient, the effective gap thickness and the gap conductance for coated surfaces.

Note: The molecular masses of commonly used metals for coating, such as chromium, nickel and zinc, are not much different from that of copper and, therefore, the accommodation coefficient will not change much when these metals are used for coating. Tin is a notable exception, but tin being a soft material and a good conductor of heat, coating by tin is likely to increase the solid spot conductance (see Chap. 6).

Misra and Nagaraju also noted that the gap conductance increased when the coating was thicker. They attributed this to the surfaces becoming smoother when the coating thickness is increased and consequent reduction in the physical gap thickness. This was confirmed by SEM analysis of the coated surfaces.

4.7 Gas Gap Conductance: Conclusions

The following conclusions follow as a result of the topics discussed in this chapter.

1. The gas gap conductance is mainly a function of the mean gap thickness and the thermal conductivity of the gas. However, other factors as listed below, might significantly affect the gap conductance.
2. The gap thickness depends on the surface roughness and reduces with the contact pressure. The consequent increase in gas conductance, however, is moderate compared with the corresponding increase in the solid spot conductance.
3. The gap conductance depends on the temperature jump distance which depends not only on the gas thermophysical properties, but also on the accommodation coefficient. At any given temperature, the accommodation coefficient depends on the gas, the solids in contact and the condition of the solid surface.
4. For a given pair of surfaces and temperature, there is a threshold gas pressure (typically, 100 mm Hg or 13 kPa) above which increase of the conductance with gas pressure is insignificant.
5. If the gas pressure is below 1 mm Hg (0.13 kPa), the gas conduction contribution to heat transfer is negligible.
6. For intermediate pressures and/or when the physical gap thickness is small, free molecular heat conduction may be important.
7. Gap conductance may also be calculated by one of the several correlations available. All of the correlations start from the fact that the gap conductance is obtained by dividing the thermal conductivity of the gas by the sum of the physical gap thickness and the temperature jump distances.
8. Recent theoretical analyses have shown that it is generally acceptable to add the gap conductance to the solid spot conductance to determine the total contact conductance.
9. Coating of surfaces with metals of high molecular mass, such as silver or gold, will result in a lower accommodation coefficient and may result in reduce gap conductance. The effect is mitigated to some extent in consequence of the surfaces becoming smoother after coating and thus reducing the effective gap thickness.

4.8 Gap Fluid is a Liquid

It is usually assumed that when the contact interface is filled with a liquid, the liquid completely wets the surfaces and fills the gaps completely. The heat transfer coefficient for the gap can be then simply estimated by dividing the fluid

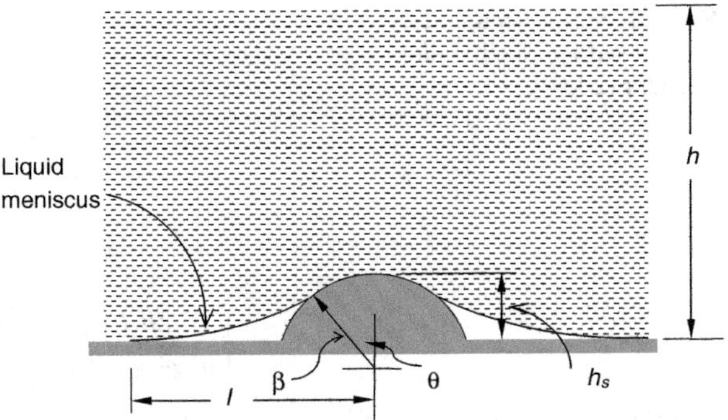

Fig. 4.6 Liquid of high surface tension in contact with a cylindrical asperity

conductivity by the mean gap thickness. This is true for the majority of liquids, such as water, which have low surface tension. For liquids with high surface tension, such as mercury, and molten metals in general, the liquid does not completely the wet surface and the real area of contact between the liquid and the solid surfaces will be less than the apparent contact area. This was explained by Timsit (1982) who analysed the contact of a liquid with high surface tension using a simple two dimensional model.

In this model, the asperities are modelled as an array of cylinders in parallel, that is, the asperities are longitudinal with radius of curvature β and height h_s, with $\beta > h_s$, as shown in Fig. 4.6.

Timsit determined the shape of the meniscus by determining the conditions required to minimise the total (gravitational plus surface) energy of the system. He used variational principles to achieve this. A capillary length, α, was defined by the equation

$$\alpha^2 = \frac{\gamma}{\rho g}$$

where γ is the surface tensionand ρ is the density of the liquid.

Using the simplifying assumptions, $h \gg hs$, $l \gg \alpha$, and that the liquid is non-adhering (meniscus profile is tangential to the surfaces in contact), the following equations were derived:

$$\theta = \left(\frac{2h_s h}{\alpha^2 + \beta h}\right)^{1/2}$$

and

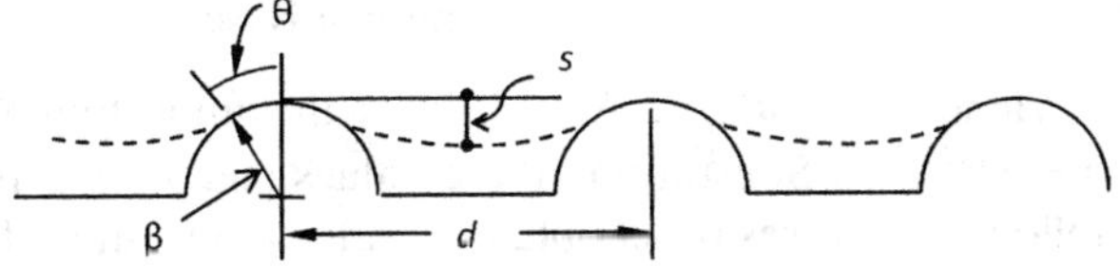

Fig. 4.7 Meniscus does not touch the valleys of the solid surface

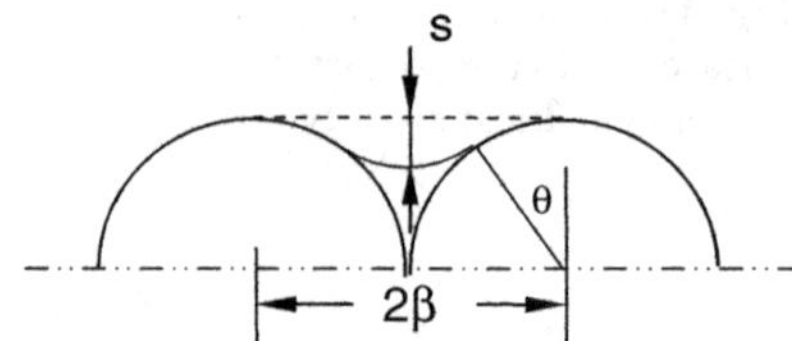

Fig. 4.8 Surface roughness model used by Heichal and Chandra (2004)

$$l = \alpha^2 \left\{ \frac{2h_s h}{h(\alpha^2 + \beta h)} \right\}^{1/2}$$

If the liquid thickness h is sufficiently small (and its surface tension high), the liquid meniscus enters the valleys between surface asperities only partially to a depth s ($< h_s$), below the tip of the asperities. Here the liquid is solely supported by the tips of the asperities as shown below Fig. 4.7:

In this case, depth of the meniscus is given by:

$$s = \frac{d^2 h}{8(\alpha^2 + \beta h)}$$

and the ratio of real to apparent contact area is:

$$f = \frac{\beta h}{\alpha^2 + \beta h}$$

It can also be from the figure seen from the figure that

$$f = \frac{2\beta sin\theta}{d}$$

On the other hand, if s = h_s (as in previous figure where the meniscus contacts the valley floor), then

$$f = 1 - \frac{2l}{d}$$

The above theory was used by Heichal and Chandra (2004) and Xue et al. (2007) in their analyses on the impact of a molten metal droplet on a solid surface. The model used by them for the surface profile was slightly different, as shown in the figure below Fig. 4.8:

It is at once seen that, in this case, $d = 2\beta$ and, therefore $f = sin\theta$. We can also see that, for the surface profile as shown, the arithmetic average roughness is

$$R_a = \frac{\text{Area of Asperities}}{\text{Length of Base}} = \frac{\pi\beta^2/2}{2\beta} = \frac{\pi\beta}{4}$$

Hence $\beta = 4R_a/\pi$. Also if V is the impacting velocity of the droplet, then $h = (V^2/2\ g)$. Substituting these values and $\alpha^2 = \gamma/\rho g$ in the expression for f, the following expression is obtained (after a few simplifications):

Table 4.9 Properties of molten metals, (Heichal and Chandra 2004)

	Tin	Bismuth	Aluminium 380
Melting point [°C]	232	271	Solidus: 538 Liquidus: 593
Thermal conductivity [W/(mK)]	32	7.9	109
Specific heat [J/(kgK)]	253	120	963
Density [kg/m^3]	6970	9750	2760
Surface tension [N/m]	0.53	0.37	0.89

Notice, however, that the thermal conductivity of aluminium alloy is 13.8 times that of bismuth; this is likely to compensate for the reduced contact area with the aluminium droplet

$$f = \frac{\frac{\rho V^2}{2}}{\frac{\rho V^2}{2} + \frac{\pi\sigma}{4R_a}}$$

Note: *The above equation may also be derived by simple equilibrium considerations.*

Since $\rho V^2/2$ = the *static pressure p*, an alternative expression for *f* will be

$$f = \frac{p}{p + \frac{\pi\sigma}{4R_a}}$$

Note that the above expressions are applicable only to the particular surface profile chosen.

The heat transfer across the interface was then calculated using the actual contact area and mean thickness of the gap R_a `assuming one-dimensional flow.

The authors noted that bismuth has much higher density and lower surface tension than aluminium alloy (see Table 4.9) and, therefore, penetrates deeper into surface asperities, producing lower contact resistance. In this connection, we may note the recent experiments of Daryabeigi et al. (2012) who were measuring the thermal conductivity of Lockheed Insulation LI-900. They were able to virtually eliminate the contact resistance between the cold side surface of the insulation sample and the cold plate by introducing a thin layer of liquid bismuth in the interface (see also Chap. 9),

References

Aaron RL (1963) A theoretical study of the thermal conductance of joints with varying ambient pressure. Southern Methodist University, Dallas, MS thesis

Antonetti VW (1992) Statistical variability of thermal interface conductance. In: NSF/DITAC workshop, Melbourne, Australia, pp. 37–45

Blodgett KB, Langmuir I (1932) The accommodation coefficient of hydrogen: a sensitive detector of surface films. Phys Rev 40:78–104

Boeschoten F, Van der Held EFM (1957) The thermal conductance of contacts between aluminium and other metals. Physica 23:37–44

Cassidy JF, Mark H (1969) Thermal contact resistance measurements at ambient pressures to 3×10^{-12} Mm Hg and comparison with theoretical predictions. NASA, Technical Memorandum X-52566

Cetinkale TN, Fishenden M (1951) Thermal conductance of metal surfaces in contact. In: Proceeding of general discussion on heat transfer. Institute of Mechanical Engineers, London, pp 271–275

Cohen I, Lustman B, Eichenberg JD (1960) Measurement of thermal conductivity of metal-clad uranium oxide rods during irradiation. Report Wapd-288. Bettis Atomic Power Laboratory, Washington, DC

Daryabeigi K, Jeffrey R, Knutsonet JR, Cunnington GR (2012) Reducing thermal contact resistance for rigid-insulation thermal measurements. J Thermophys. Heat Transfer 26:172–175

Dharmadurai G (1983) Estimation of thermal accommodation coefficients for inert gases on nuclear materials. J App Phys 54(10):5990–5992

Dutkiewicz RK (1966) Interfacial gas gap for heat transfer between two randomly rough surfaces. Third Int Heat Transfer Conf 4:118–126

Hall ROA, Martin DG, Mortimer MJ (1990) The thermal conductivity of UO_2 sphere-pac beds. J Nucl Mat 173:130–141

Hardee HC, Green WR (1968) Thermal conductivity in small air gaps at very low pressures. Sc-Tm-68–309, Sandia Laboratory, Albuquerque, NM

Heichal Y, Chandra S (2004) Predicting thermal contact resistance between molten metal droplets and a solid surface. Trans ASME J Heat Transfer 127:1269–1275

Kennard EH (1938) Kinetic theory of gases. McGraw-Hill, New York, pp 311–327

Kharitonov VV, Kokorev LS, Del'vin NN (1973) On the role of the accommodation coefficient in contact heat exchange. Atomnaya Energiya 35(5):360–361

Knudsen M (1934) The kinetic theory of gases. Methuen, London, pp 46–61

Lang PM (1962) Calculating heat transfer across small gas-filled gaps. Nucleonics 20(l):62–63

Madhusudana CV (1975) The effect of interface fluid on thermal contact conductance. Int J Heat Mass Transfer 18:989–991

Madhusudana CV (1993) Thermal contact conductance and rectification at low joint pressures. Int Comm Heat Mass Transfer 20:123–132

Madhusudana CV, Fletcher LS (1981) Gas conductance contribution to contact heat transfer. AIAA Paper 81–1163, American Institute of Aeronautics and Astronautics, New York

Majumdar A, Williamson M (1990) Effect of interstitial media on contact conductance: a fractal approach. ASME HTD 153:49–57

Misra M, Nagaraju J (2010) Thermal gap conductance at low contact pressures (< 1 MPa): Effect of gold plating and plating thickness. Int J Heat and Mass Transfer 53:5373–5379

Negus KJ, Yovanovich MM (1988) Correlation of the gap conductance integral for conforming rough surfaces. J Thermophy Heat Transfer 2:279–281

Nishino K, Torii K (1994) Thermal contact conductance of wavy metal surfaces under arbitrary ambient pressure. In: Proc 10th Intl Heat Transfer Conf, Paper No. 15 C-118. Institution of Chemical Engineers, Rugby, UK

Popov VM, Krasnoborod'ko AI (1975) Thermal contact resistance in a gaseous medium. Inzhenerno-Fizicheski Zhurnal 28(5):875–883

Rapier AC, Jones TM, McIntosh JE (1963) The thermal conductance of uranium dioxide/stainless steel interfaces. Int J Heat Mass Transfer 6:397–416

Roberts JK (1932) The exchange of energy between gas atoms and solid surfaces ii: The temperature variation of accommodation coefficient of helium. Proc R Soc Lond Ser A 135:192–205

Semyonov YuG, Borisov SE, Suetin PE (1984) Investigation of heat transfer in rarefied gases over a wide range of knudsen numbers. Int J Heat Mass Transfer 27:1789–1799

Shaikh RA, Beall AN, Razani A (2001) Thermal modeling of the effect of interstitial fluid on contact resistance between two dissimilar materials. Heat Transfer Eng 22:41–50

Shlykov YP, Ganin YA (1964) Thermal resistance of metallic contacts. Int J Heat Mass Transfer 7:921–929

Smoluchowski MS (1898) On conduction of heat in rarefied gases. Phil Mag J Sci Ser 5:192–206

Song S, Yovanovich MM (1987) Correlation of thermal accommodation coefficient for engineering surfaces. ASME HTD 69:107–116

Song S, Yovanovich MM, Nho K (1992) Thermal gap conductance: effects of gas pressure and mechanical load. J Thermophys Heat Transfer 6:62–68

Song S, Yovanovich MM, Goodman FO (1993) Thermal gap conductance of conforming rough surfaces in contact. Trans ASME J Heat Transfer 115:533–540

Thomas LB, Brown RE (1950) The accommodation coefficient of gases on platinum as a function of pressure. J Chem Phys 18:1367–1372

Thomas LB, Loyalka SK (1982) Determination of thermal accommodation coefficients of inert gases on a surface of vitreous UO_2 at 35 °C. Nucl Technol 59:63–69

Timsit RS (1982) The true area of contact between a liquid and a rough solid: elementary considerations. Wear 83:129–141

Tsukizoe, T and Hisakado, T (1965) On the mechanism of contact between metal surfaces – the penetrating depth and the average clearance, Trans ASME, J Basic Eng, 87:666–674

Ullman A, Acharya R, Olander DR (1974) Thermal accommodation coefficients of inert gases on stainless steel and UO_2. J Nucl Mat 51:277–279

Vickerman RH, Harris R (1975) The thermal conductivity and temp jump distance of gas mixtures. American Nuclear Society, winter meeting. American Nuclear Society, San Francisco

Wachman HY (1962) The thermal accommodation coefficient: a critical survey. ARS J 32:2–12

Wahid SMS, Madhusudana CV (2000) Gap conductance in contact heat transfer. Int J Heat Mass Transfer 43:4483–4487

Wiedmann ML, Trumpler PR (1946) Thermal accommodation coefficients. Trans ASME 68:57–64

Xue M, Heichal Y, Chandra S, Mostaghimi J (2007) Modeling the impact of a molten metal droplet on a solid surface using variable interfacial thermal contact resistance. J Mater Sci 42:9–18

Yovanovich MM Marotta EE (2003) Thermal spreading and contact resistances. In:Bejan A, Kraus AD (eds.) heat transfer Handbook, Wiley, New York

Yovanovich MM, DeVaal JW, Hegazy AA (1982) A statistical model to predict thermal gap conductance between conforming rough surfaces. AIAA Paper 82-0888. American Institute of Aeronautics and Astronautics, New York

Chapter 5
Experimental Aspects

Thermal conductance of joints may be determined experimentally in several ways. However, by far the most common method uses the axial heat flow apparatus, based on the method described in ASTM E1225–09 in which the two cylinders of similar or dissimilar materials are placed end to end as illustrated in Fig. 1.2 (Chap. 1). There have been other apparatus built for specific needs, for example, to determine the contact conductance in duplex tubes when the heat flow is radial; in periodic contacts and in manufacturing processes. Also used frequently are transient heat flow measurements to establish thermal contact properties. The relative merits of steady state and transient methods are also discussed in this chapter.

In every case, before the heat transfer experiments are performed, it is necessary that profilometric measurements are made to characterise the contacting surfaces. It is also desirable to determine the microhardness or similar properties of the surfaces prior to the heat transfer tests.

Apart from heat transfer apparatus, conducting sheet and electrolytic tank analogues have been constructed and used, mainly to determine the resistance of various shapes of constrictions. Some of these equipment will also be briefly described in this chapter.

The chapter also includes a discussion of the accuracy of measurements as applicable to contact heat transfer.

5.1 Axial Heat Flow Apparatus

This apparatus is based on the method described in ASTM E1225–09 for thermal conductivity measurements using longitudinal heat flow. Several investigators have used this type of experimental rig for contact conductance measurements, for example, Cetinkale and Fishenden (1951), Williams (1966), Mikic and Rohsenow (1966), Fletcher et al. (1969) and O'Callaghan and Probert (1972).

A schematic of a typical axial heat flow apparatus is shown in Fig. 5.1. Essentially, the rig consists of two cylinders placed end to end and loaded in the axial direction either mechanically (see Fig. 5.2) or by hydraulic means. If the

C. V. Madhusudana, *Thermal Contact Conductance*,
Mechanical Engineering Series, DOI: 10.1007/978-3-319-01276-6_5,

loading is achieved by other than mechanical means, then the contact load needs to be measured using a calibrated load cell. Axial heat flow is achieved by providing a heat source at the top end and a heat sink at the bottom end. Frequently, however, provision for heating as well as cooling of either end is made so that the effect of reversing the heat flow direction may be studied, without dismantling and re-assembling the specimens. Generally, the assembly is placed in a chamber that can be evacuated in order that the solid spot conductance may be isolated and determined. To transmit the mechanical load to the assembly inside the chamber, a bellows or similar device would be required. A rotary vane type vacuum pump is satisfactory for pressures down to about 10^{-3} Torr (0.133 Pa). A vacuum diffusion pump will also be needed if lower pressures of the order of 10^{-6} Torr are required. The traditional method of heating was by the use of electrical resistance coils such as Nichrome and cooling by water circulating in a coil and supplied by a constant head tank. Constant temperature circulating baths are also commonly used to provide both heating and cooling.

Heat flux meters, made of certified standard reference materials (SRM's), measure the heat flow through the specimens. They also facilitate in situ determination of the thermal conductivities of the test bars, thus eliminating the need to refer to external sources for this information.

To minimize heat losses, a guard heater or a thermal shield should be placed around the test section. This is especially important if the tests are done in a conducting medium such as air. In addition, insulators need to be placed at each end of the column assembly to prevent heat losses in the axial direction.

The usual method of measuring the temperatures along the axes of the specimens is by means of calibrated thermocouples, although other highly accurate and

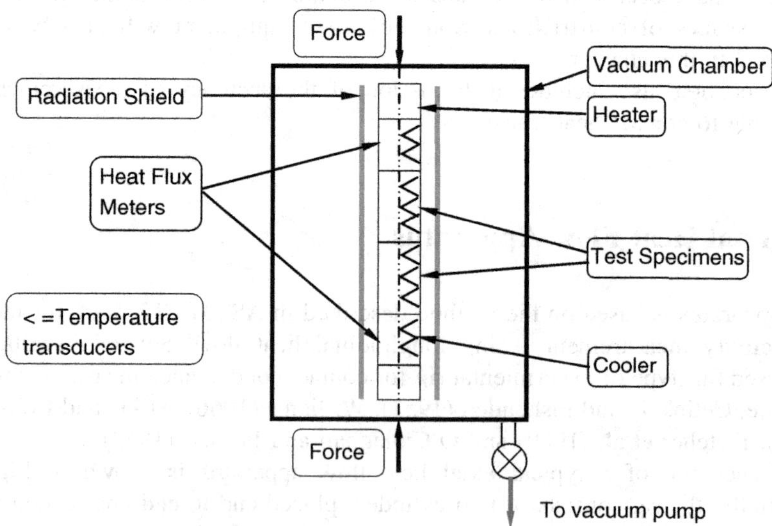

Fig. 5.1 Schematic diagram of the axial heat flow apparatus

Fig. 5.2 An axial heat flow apparatus showing the mechanical loading arrangement by lever and dead-weights (Villanueva 1997)

sensitive temperature sensing devices such as thermistors and resistance temperature detectors (RTD's) are frequently used. The thermocouples are inserted in holes drilled normal to the axis and extending to the axis, that is, the length of the holes is equal to the radius of the specimen. The use of a conducting cement or a soft foil is necessary at the bottom of the hole to provide good contact between the thermocouple and the specimen. The output from the temperature sensors is usually fed into computer based data takers for immediate processing of experimental data.

Before we leave the section on axial heat flow apparatus, it is not out of place to mention one of the early experimental works (Phillips 1956) on heat transfer across joints placed in different gaseous atmospheres. A noteworthy feature of this work is to enclose just the joint in a cylindrical container or "sealing gasket", rather than place the whole assembly in a vacuum tight chamber.

The sealing gasket was machined from a solid cylinder of Teflon, four and one-half inches (114 mm) in diameter and six inches (152 mm) long. The gasket was machined 0.010 inch (2.54 mm) undersize with respect to the specimen. The gasket material was drilled and tapped to receive a 3/16 inch (4.76 mm) stainless steel nipple through which the various gaseous media were introduced. An adjustable stainless steel banding clamp was fitted over the Teflon gasket after it

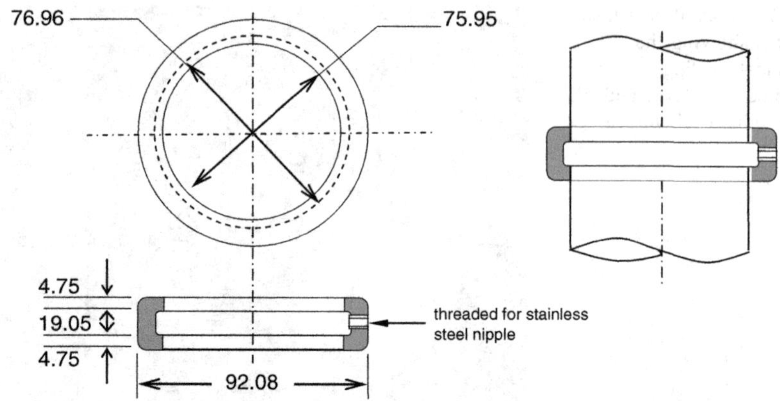

Fig. 5.3 Sealing gasket to maintain the interface in vacuum or in gaseous atmosphere

was in place on the test specimens in order to hold it more securely in place and to better enable it to hold its pressure and/or vacuum after thermal expansion. A sketch of the sealing gasket is shown in Fig. 5.3 (dimensions in mm).

Phillips was able to test the interfaces in vacuum, air and helium with the aid of this gasket alone. No vacuum (or pressure tight) chamber was needed. His results were consistent with what were reported for these conditions in later decades. If this method could be successfully applied, it would considerably simplify the design, construction and operation of axial heat flow apparatus. Of course, the lateral heat loss from the specimens should be minimised by the use of either good quality insulation (Kempers et al. 2009) or by the provision of guard heaters.

5.2 Radial Heat Flow Apparatus

In a cylindrical joint transmitting heat radially, the contact pressure is developed (or the initial shrink-fit pressure is modified) as a result of the differential expansion of the two cylinders. As will be seen in a later chapter, the differential expansion and, hence, the contact pressure are functions of the heat flux. To test these types of joints, therefore, no provision needs to be made for mechanical loading (see Cohen et al. 1960; Williams and Madhusudana 1970; Hsu and Tam 1979; Madhusudana and Litvak 1990). On the other hand it is important to be able to measure the heat flux accurately.

The essentials of a typical radial heat flow apparatus is shown in Fig. 5.4. A major problem in testing of this type of joint appears to be the difficulty in obtaining a truly axisymmetric temperature distribution in the specimens if the inside surface of the inner cylinder is heated by means of a central non-contacting rod. The use of a pre-heated liquid, instead, may alleviate the difficulty to some extent; however, this method is going to cause additional problems if testing in vacuum is required.

A second source of problems arises because of the need to make the cylinders sufficiently long to avoid the end effects; firstly, very accurate machining to close tolerances is required to ensure that the ends are straight and parallel with negligible taper; secondly, extra care is required in the assembly of long cylinders to produce a shrink fit; thirdly, it will be necessary to drill deep holes of very small diameter to locate the thermocouples inside the wall thickness of the specimens. In the apparatus shown in Fig. 5.2, temperatures were measured only on the surfaces of the specimens as it would have been impossible to locate sufficient number of thermocouples along the radial co-ordinate to measure the temperature distribution reliably. Perhaps because of these difficulties, there has been comparatively small number of reports dealing with the experimental investigation of radial heat flow in composite cylinders.

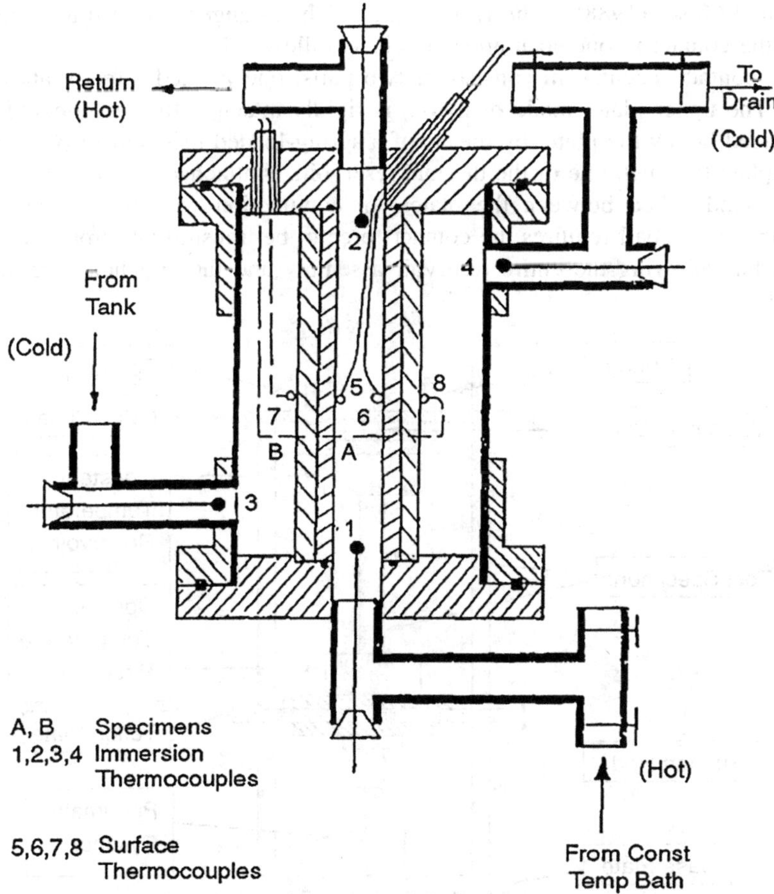

Fig. 5.4 Radial heat flow apparatus (Madhusudana and Litvak 1990)

An associated problem is the determination of TCR in finned tubes. Experimental apparatus specially devised for this application (Cheng and Madhusudana 2006) is described in a later chapter.

5.3 Periodic Contacts

There are several applications in which the heat transfer occurs between surfaces that are undergoing a regular cycle of contact and separation. Examples include the heat transfer between the exhaust valve and its seat in an internal combustion engine, the heat transfer between the die and the work piece in a repetitive hot metal deformation process.

Some experimental apparatus used to test periodic contacts is similar to the axial heat flow facility described earlier. The experimental apparatus designed by Dodd and Moses (1988) is shown in Fig. 5.5. The arrangement used to make and break the contact in one such apparatus is as follows.

The contact mechanism consists of two parts, one located directly above the other. The upper plate, made of nylon, is rigidly attached to the support frame. Suspended below the plate, by means of a spring-loaded mechanism is a smaller nylon plate to which one of the thermal reservoirs is attached. A nylon alignment ball is sandwiched between the plates. In conjunction with the spring-loaded mechanism, the ball requires the contact force to be transmitted through a single point, thus allowing the entire reservoir assembly and the attached specimen to

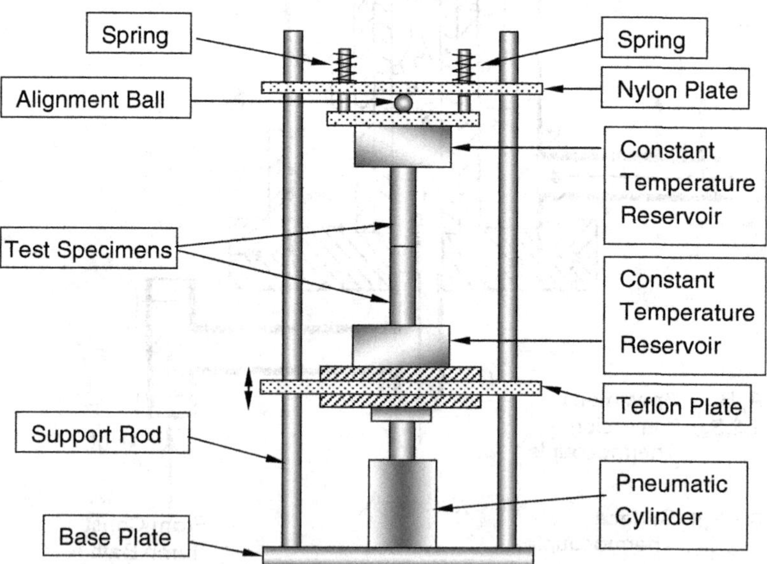

Fig. 5.5 Experimental apparatus for periodic contacts (after Dodd and Moses 1988)

pivot about that point, bringing the surfaces of the specimens into contact in their entirety. The second test specimen and its associated thermal reservoir is attached to the lower plate, made of Teflon, which is free to slide along the four PVC rods forming the support frame.

The test specimens were caused to contact and separate by driving the lower plate by means of a pneumatic cylinder. The airflow to the cylinder was controlled by a dual acting solenoid valve, with one air stream used to drive the test specimen into contact and the other to separate the two at the end of the contact period of the cycle. The valve was micro-processor controlled and the data was processed by the use of a Data Acquisition and Control unit.

In an earlier investigation by Howard (1976) into periodic contacts, it was the upper bar that was moved by actuating the pneumatic cylinder. Both apparatus allowed tests to be conducted in ambient atmosphere only—there was no provision for evacuating the test section.

5.4 Transient Measurements

The advantages claimed for the steady state method are simplicity, direct measurements without having to rely on "handbook" data for properties such as the thermal conductivities of test materials, accuracy and simple calculations required to estimate the conductance. The main disadvantage of steady-state methods is the long time it takes to perform each set of experiments. On the other hand, results are obtained relatively quickly in transient measurements. In general, the thermal contact resistance is determined by matching the measured temperature–time history with the numerically predicted one by means of fitting parameters. However, these techniques invariably require independent measurements of such thermophysical properties as the specific heat or the thermal conductivity of the sample materials. Further, the conductance values determined are time-dependent and corrections need to be applied to them before they are used in thermal design of apparatus. These factors introduce additional uncertainties.

Transient numerical methods are necessarily implicit. The measured variation of the temperature with time in one or both of the specimens in contact is compared against that obtained from a numerical (or analytical) approach with an assumed value of the thermal resistance. The thermal resistance in the numerical model is then adjusted until the computed values agree with the measured values.

5.4.1 Transient Methods

In some experimental methods using the transient approach, the two surfaces are initially in contact at the same temperature. At the start of the experiment, a temperature disturbance is introduced by heating and/or cooling one of the specimens. Bosch and Lasance (2000), for example, describe a facility in which a sample is sandwiched between a copper upper cylinder and an aluminium support. There are thus two interfaces: copper/sample and sample/aluminium (Fig. 5.6). The aluminium support can be heated or, alternatively, maintained at ambient temperature by connecting it to one of two water baths. The assembly is initially in thermal equilibrium at ambient temperature. The hot water is switched on for about 30 s followed by a switch back to ambient temperature. The temperature–time histories of the copper block and the aluminium support are recorded. These are then compared with the results of a *numerical simulation* using the total resistance, R_{th} defined below, as the fitting parameter.

$$R_{th} = d/(kA) + 2R_c$$

In this expression:
k = thermal conductivity of the sample
A = area of sample normal to heat flow
d = thickness of the sample
R_c = thermal contact resistance of either interface
Literature values of the thermal conductivity were used in evaluating the above equation.

In the *laser-flash method* for measuring the thermal diffusivity of a material, a small disc-shaped sample (about the size of a small coin) is subjected to a very short burst of radiant energy in the form of the laser-flash. The duration of the irradiation time is one millisecond or less. The resulting temperature rise of the rear surface of the sample is measured and thermal diffusivity values are computed from the temperature rise versus time data (Fig. 5.7).

The theory behind the method is as follows (Hohenauer 1999):

For one dimensional flow of heat

$$\frac{\partial^2 T}{\partial x^2} = \frac{1}{\alpha}\left(\frac{\partial T}{\partial t}\right) \tag{5.1}$$

Solving by separation of variables

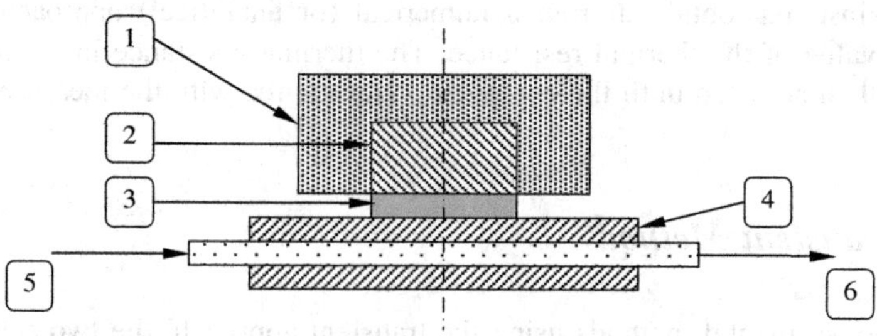

Fig. 5.6 Transient heat flow apparatus. *1* Insulation *2* Copper block *3* Specimen *4* Aluminium block *5* Water inlet *6* Water outlet

$$T(x,t) = \left[A\sin\sqrt{\frac{c}{\alpha}}x + B\cos\sqrt{\frac{c}{\alpha}}x\right]e^{-ct} \tag{5.2}$$

For no heat loss from the surface at $x = 0$, we must have $A = 0$. The second boundary condition, $\frac{\partial T}{\partial x} = 0$ at x = d (where d is the thickness of the sample) gives

$$c = \frac{n^2\pi^2\alpha}{d^2}$$

Therefore, the solution will be of the form

$$T(x,t) = \sum_{n=0}^{\infty}\left[B_n cos\frac{n\pi x}{d}\right]e^{-\frac{n^2\pi^2\alpha t}{d^2}} \tag{5.3}$$

For t = 0, this equation may be rewritten as

$$T(x,0) = B_0 + \sum_{n=1}^{\infty}\left[B_n cos\frac{n\pi x}{d}\right] \tag{5.3a}$$

The heat pulse, applied when t = 0, is of infinitesimal duration, With this condition, B_O and B_n may be evaluated as: $B_0 = T_0 + \Delta T_{max}$ and $B_n = 2\Delta T_{max}$ where T_0 is the initial temperature and T_{max} is the temperature reached by the entire specimen after a sufficiently long temperature; $\Delta T_{max} = T_{max} - T_0$. Hence

$$T(x,t) = T_0 + \Delta T_{max}\left[1 + 2.\sum_{n=1}^{\infty} cos\frac{n\pi x}{d}e^{-\frac{n^2\pi^2\alpha t}{d^2}}\right] \tag{5.4}$$

and

$$T(d,t) = T_0 + T_{max}\left[1 + 2.\sum_{n=1}^{\infty}(-1)^n e^{-\frac{n^2\pi^2\alpha t}{d^2}}\right] \tag{5.4a}$$

Using the *first order approximation*, the half-time, that is, the time required for the rise of temperature to reach *half* its maximum value, is then given by

$$\frac{\Delta T}{\Delta T_{max}} = \frac{1}{2} = \left[1 + 2(-1)e^{-\frac{n^2\pi^2\alpha t}{d^2}}\right] \tag{5.5}$$

This gives

$$\frac{1}{4} = \left[e^{-\frac{\pi^2\alpha t_{0.5}}{d^2}}\right] \tag{5.6}$$

Taking logarithms and solving for α,

$$\alpha = -\frac{ln\left(\frac{1}{4}\right)}{\pi^2}\left(\frac{d^2}{t_{0.5}}\right) = 0.140\left(\frac{d^2}{t_{0.5}}\right) \tag{5.7}$$

Thus, the diffusivity can be determined once the half-time, $t_{0.5}$, is estimated from the graph (Fig. 5.7).

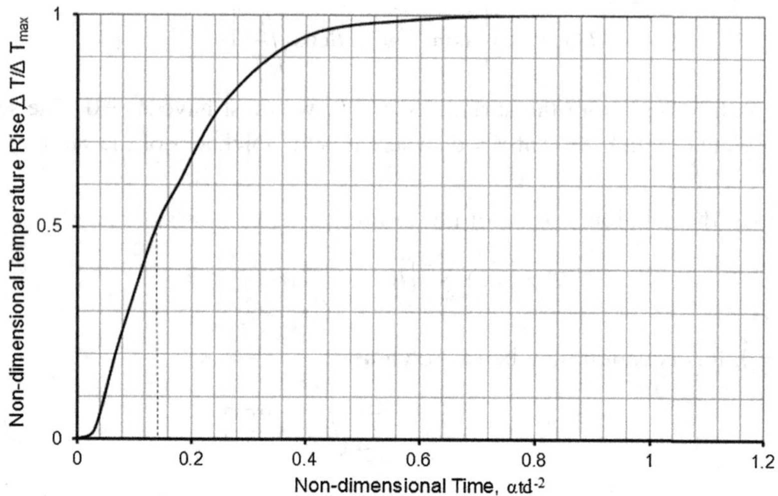

Fig. 5.7 Transient temperature–time graph showing half-time

Note The above analysis is based on simplifying assumptions. Recent software for the analysis of results from the laser flash apparatus allows for non-adiabatic boundary conditions and energy pulses of finite duration.

When the laser-flash apparatus is used to determine thermal contact conductance, a fitting procedure is normally used. A series of temperature–time graphs are first obtained (by numerical analysis) for the sample with an interface, for various assumed values of TCC. The TCC corresponding to that graph which matches experimentally measured temperature–time graph is taken to be the TCC of the sample.

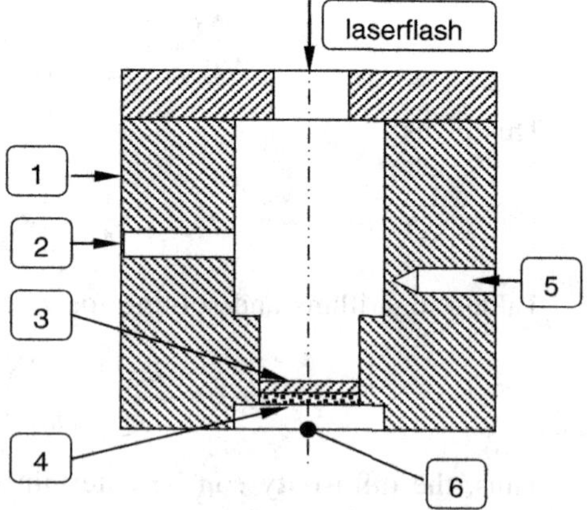

Fig. 5.8 Part of a laser flash apparatus. *1* Pressure chamber. *2* Gas inlet/outlet. *3* Copper plate. *4* Specimen. *5* Thermocouple well. *6* MgF_2 window

Griesinger et al. (1998) used the laser-flash apparatus for the measurement of the TCC between a solid (copper) surface and zeolite powder (Fig. 5.8). In this case, the duration of the pulse is of the order of 1 s or less. *Again, the TCC is determined by an inverse procedure*. The numerically calculated temperature rise at the back of the specimen (zeolite layer) is matched with the measured temperature rise using (assumed values of) the TCC, as well as the thermal diffusivities of copper and zeolite as the fitting parameters. The tests could be conducted in chamber pressures from 0.0015 bar (150 Pa) to 1 bar (100 kPa).

Although accuracies of 1 % are claimed for the diffusivity of single material samples, the uncertainties associated with laser-flash measurements of TCC are not yet established. Greisinger et al. claim that the results are reproducible to +10 %.

The experimental approach presented by Fieberg and Kneer (2008) was based on transient infrared temperature measurements. Two bodies initially at two different temperatures were brought in contact and the surface temperature histories were recorded with a high-speed infrared camera. The contact heat flux was calculated by solving the related inverse problem. From the contact heat flux and from the measured temperature jump at the interface the contact heat transfer coefficient was calculated.

The inverse method used for the calculation of the heat flux was based on the analytical solution for a semi-infinite body and a step response to a Neumann (heat flux) boundary condition.

Readers interested in theoretical analysis may like to refer to the analytical solution for determining the TCR between the materials of a double layer sample using laser flash method presented by Milosevic et al. (2002).

As the name implies, the transient contact conductance is a function of time. It is, therefore, sometimes claimed that the use of steady state conductance values for transient conditions will result in a large error. It is interesting, therefore, to consider how the TCC varies with time.

An original work analyzing the variation of the *constriction resistance* with respect to time is that of Schneider et al. (1977). As we have seen earlier, the constriction resistance is the resistance associated with a *single* contact spot of radius a and TCC is the sum of the reciprocals of all of the constriction resistances in the contact plane. The results of their numerical analyses for copper/stainless steel, copper/glass, copper/steel and steel/glass interfaces could be correlated into a single equation:

$$\frac{R_{tr}}{R_{ss}} = 0.43\tanh[0.37ln(4X)] + 0.57 \quad (5.8)$$

where

R_{ss} = steady state constriction resistance
R_{tr} = transient constriction resistance
X = correlation parameter $= 0.5\left[1 + \sqrt{\alpha_2/\alpha_1}\right] F_{o,\alpha}$

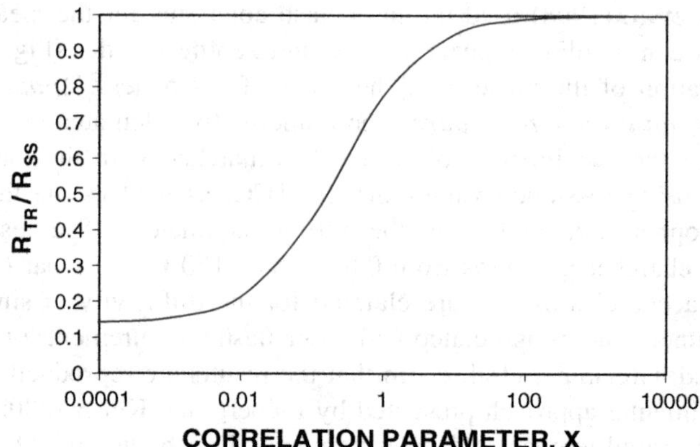

Fig. 5.9 Variation of disc constriction resistance with time [Plotted from Eq. (5.8), Schneider et al. 1977]

Table 5.1 Time required for the constriction resistance to reach the steady state value

Material 1/material 2	α_1 (m^2/s) $\times$ 10^5	α_2 (m^2/s) $\times$ 10^5	$F_{o\alpha}$ for steady state	Time for steady state, seconds
Copper/steel	13.2	1.36	151.40	0.0055
Steel/glass	1.36	0.06	165.28	0.1294
Copper/stainless steel	13.2	0.49	167.69	0.0160
Copper/glass	13.2	0.06	187.37	0.1412

$\alpha_m = 2\alpha_1\alpha_2/(\alpha_1 + \alpha_2)$.
$F_{o,\alpha} = \alpha_m t/a^2$.
α = thermal diffusivity, ρ/kC
ρ = density
C = specific heat
The results are shown plotted in Fig. 5.9
With X as defined above, we see that the Fourier Number is

$$F_{o,a} = \frac{2X}{1 + \sqrt{\alpha_2/\alpha_1}}$$

Therefore, for a given pair of surfaces $F_{o,a}$ is directly proportional to X. Consequently X is a measure of the time elapsed. From Fig. 5.7, we see that the full, or steady state, value of the constriction is reached when $X \approx 100$. The average contact spot radius is 30 μm. The Table 5.1 may, therefore, be constructed.

Since the time taken to reach the steady state value is only a fraction of a second, it is unlikely that an error is introduced on this basis. This discussion applies only to the constriction resistance. In the actual experiment, of course, allowance must be made for the thermal capacities of the solid blocks and the specimen. It is equally important, of course, that the time constants for the measurement system, including the temperature sensors, be as small as practicable in order that the temperature–time history is faithfully recorded. Alternatively, appropriate corrections must be applied to the measurements.

5.4.2 Transient Measurements: Conclusions

As a result of the above discussion of transient measurement techniques, and the comparison with the steady state procedures, we can list the following conclusions:

- Steady-state method allows direct measurements to be made without the need for further measurements in separate apparatus to determine other required thermal properties.
- The transient methods are implicit and, in general, depend on matching the results of numerical analysis with the measurements by means of fitting parameters. (Note that built-in sophisticated software is available with modern laser flash apparatus, eliminating the need for separate analysis).
- These methods also depend on separate thermal property measurements or, alternatively, data from literature, for final results. For these reasons, there is inherently more uncertainty in the results of transient measurements.
- While the constriction resistance itself reaches the steady-state value very quickly, in the actual experiment allowance must be made for the thermal capacities of the solids on either side of the sample and the sample material itself. Thus, care must be taken to see that steady-state values are not used for transient conditions.
- For transient measurements, the time constants for the measuring system need to be as small as practicable. Corrections should be applied if necessary.
- The main disadvantage of the steady-state measurements is the relatively large time taken for each set of data to be reproduced. For this reason, transient methods may be used if the time available is short.

Transient methods, of course, are also necessary when the contact process itself is transient, e.g., intermittent contacts, manufacturing processes and dry friction. A comparison of steady state and transient techniques was presented by Madhusudana and Garimella (2003).

5.5 Analogue Methods

With the wide availability of dedicated software and consequent ease to model systems of complex geometry, the analogue methods, in general, have gone out of favour in the solution of heat transfer problems. Nevertheless, the a brief discussion is included

a. to provide a historical perspective to contact heat transfer research

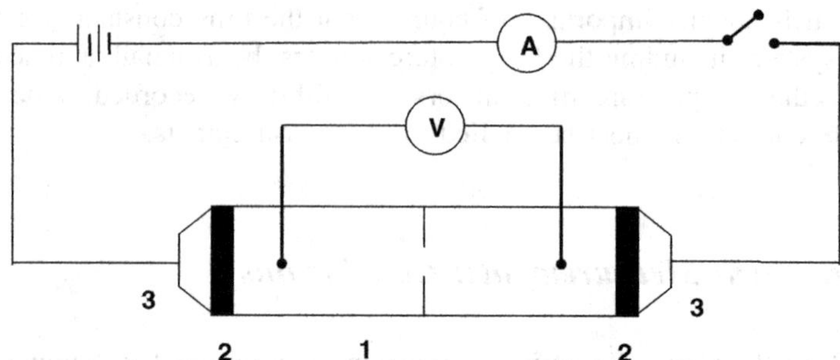

Fig. 5.10 Conducting sheet analogue for two-dimensional constrictions. *1* Conducting sheet. *2* Silver paint. *3* Clips

b. so that the similarity between heat transfer and other physical phenomena may be appreciated.

The analogue method is often a quick and inexpensive way of obtaining solutions to potential flow problems. In the current context, this method depends on the similarity between the electric voltage and the temperature since both these potentials obey the same (Laplace) equation. In contact resistance work, the analogue approach is used mainly to determine constriction resistances of various types.

For two-dimensional problems in the x−y plane, the problem is easily simulated. The heat flow region is simulated by an electrically conducting sheet ("teledeltos" paper). Prescribed voltages are applied to the silver painted ends of the paper to simulate isothermal boundary conditions (see, for example, Veziroglu and Chandra 1969). The schematic of the set-up is shown in Fig. 5.10. A cut is made in the middle of sheet to avoid any electrical contact across the cut.

The resistance with and without the cut is measured, from which the additional resistance due to constriction is calculated by the difference. Note that, in this method, it is also easy to obtain the equipotential lines (isotherms) by the use of standard equipment such as Servomex Field Plotter. The analogy has also been used in the determination of fin conduction shape factors required in the analysis of finned-tube heat exchangers (Sheffield et al. 1987).

For three-dimensional problems, it is necessary to use an electrolytic tank analogue (Karplus 1958). This type of equipment has been used to measure the resistance of:

1. Single constrictions of various shapes (Major and Williams 1977; Madhusudana 1992)
2. Single and multiple constrictions of circular shape (Jeng 1967; Yip and Venart 1968; Cooper 1969)
3. Macroscopic resistance in a bolted joint (Fletcher et al. 1989).

A diagram of the electrolytic cell used to determine the resistance of a single constriction is shown in Fig. 5.11. The following points must be noted in the design and use of an electrolytic tank analogue:

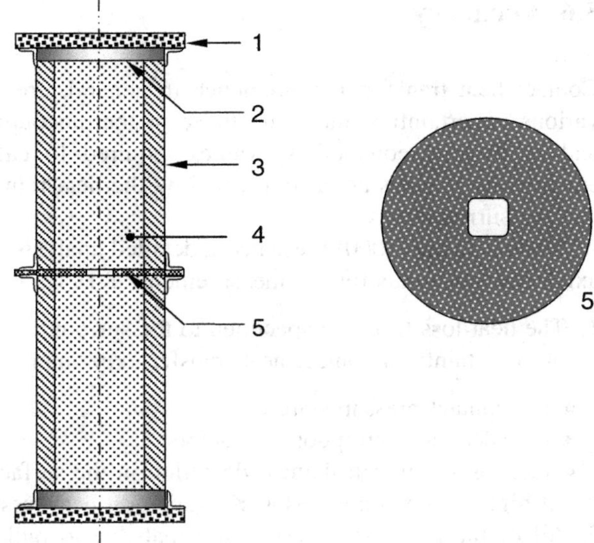

Fig. 5.11 Electrolytic tank analogue for three dimensional constrictions. *1* Perspex plate. *2* Brass electrode (Graphited). *3* PVC Pipe. *4* Electrolyte. *5* Constriction sheet

1. A container is fabricated in such a way that the shape of the electrolyte within the container is a scale model of the field configuration. Boundaries, which are equipotential, are made of metal while insulating materials or employed for adiabatic boundaries.
2. To avoid polarisation, AC voltages (frequency range 50–1,500 Hz) should be used.
3. The electrolyte must be purely resistive.
4. To minimise errors, the resistivity of the electrodes must be small compared to the resistivity of the electrolyte.
5. The surface impedance of the electrodes must be minimised by the use of graphite or platinum black coatings.

If the measurement of the resistance is the only requirement, then the method is quite simple. The constriction is simulated by a plastic sheet cut to the required shape. Two measurements of the conductivity of the cell is made—one with the constriction in place and the other without the constriction—by means of an AC conductivity meter. Alternatively, the resistance may be measured in a universal bridge. Knowing the dimensions of the tank, the additional resistance due to constriction may be calculated.

If the temperature profile is also required, then a device such as the electronic-analogue field mapper, as described by Karplus (1958), may be used. However, since we are mainly interested in determining the resistances, the additional complexities involved will defeat the purpose of obtaining quick and simple solutions. Furthermore, if an accurate temperature is required, then it is desirable to perform a numerical analysis.

5.6 Accuracy

Contact heat transfer measurements, in general, are subject to error because of various uncertainties, including those in thermocouple calibration and location, and in thermal conductivity values required to calculate the heat flux. The experimental values are also affected by the heat transfer between the specimens and the surroundings.

Madhusudana (2000) presented a detailed analysis of possible heat losses in an axial flow apparatus for the measurement of TCC. The major conclusions were:

1. The heat loss from the specimen to the surroundings represents a major source of uncertainty in contact heat transfer experiments especially when:
   - the contact pressures are low
   - the specimens are poor conductors
   - there is significant flatness deviation in the surfaces of contact
2. At high temperatures (>450 K), radiation heat loss becomes significant.
3. All of the heat losses, especially that due to radiation, may be considerably reduced by the provision of a radiation shield.
4. Unless the contact pressure is very low or the flatness deviation is large, moderate vacuums of the order of 10^{-2} Torr (1.33 Pa) are satisfactory when we consider other sources of inaccuracy such as that in temperature measurement.

In an axial heat flow apparatus, with careful design and experimentation, an experimental uncertainty of <10 % is achievable for tests conducted in vacuum. For tests conducted in a conducting medium such as air, an uncertainty of 15 % is probably more representative of the accuracy to be expected. The uncertainty, especially in a conducting environment, can be reduced by the provision of carefully controlled guard heaters.

In radial heat flow apparatus, because of the difficulties in obtaining truly axi-symmetric heat distribution, in locating thermocouples accurately, and because of end effects, it is unrealistic to expect an accuracy of better than 20 % when tests are done in a conducting medium. A higher accuracy could be obtained for tests in vacuum.

Transient methods, in general, depend on separate thermal property measurements or, alternatively, data from literature, for final results. For transient measurements, the time constants for the measuring system need to be as small as practicable The transient methods are implicit and, in general, depend on matching the results of numerical analysis with the measurements by means of fitting parameters. All of these factors contribute toward uncertainties in the results of transient measurements. According to the authors quoted earlier in this chapter, accuracies of the order of 10 % are still achievable with transient techniques. With analogue methods, the uncertainty is controlled mainly by the accuracy of manufacture of the apparatus and the specimens, and the voltage measurement. With careful experimentation, uncertainties of <5 % may be reasonably accepted in both the conducting sheet and the electrolytic tank analogues. The analogue methods, however, have been mainly useful in the analysis of constriction resistance.

It is worth noting that, the very nature of contact resistance, depending as it does on the surface topography, and material properties, introduces an uncertainty that requires tests to be performed on several pairs of similarly prepared specimens to obtain reliable estimates. No such requirement is necessary for surfaces that are isotropic and random (e.g., lapped and bead blasted flat surfaces). Finally, it should be pointed out that the theoretically predicted values are also subject to error due to their dependency on measured values of (variable) microhardness and the surface profiles.

5.7 Summary

The foregoing is just a brief description of the main features of the more common types of equipment used in the experimental determination of TCR in different situations. Space does not permit a more exhaustive discussion of design considerations and details of instrumentation. Because of the rapid progress in technology and, in particular, microprocessor-based measurement and control, it is felt any such detail would have limited value. Readers interested in more information, however, may consult the references listed at the end of this chapter.

References

Bosch EGT, Lasance CJM (2000) Accurate measurement of interface thermal resistance by means of a transient method, sixteenth IEEE SEMI-THERM™ symposium, pp 167–173

Cetinkale TN, Fishenden M (1951) Thermal conductance of metal surfaces in contact. In: Proceeding of General Disc on Heat Transfer. Institute of Mechanical Engineers, London, pp 271–275

Cheng W, Madhusudana CV (2006) Effect of electroplating on the thermal conductance of fin-tube interface. Appl Therm Eng 26:2119–2131

Cohen I, Lustman B, Eichenberg JD (1960) Measurement of thermal conductivity of metal-clad uranium oxide rods during irradiation. Report Wapd-228. Bettis Atomic Power Laboratory, Pittsburgh, pp 49

Cooper MG (1969) A note on electrolytic tank analogue experiments for thermal contact resistance. Int J Heat Mass Transfer 12:1715–1717

Dodd NC, Moses WM (1988) Heat transfer across aluminum/stainless steel surfaces in periodic contact. American Institute of Aeronautics and Astronautics (AIAA), Washington, DC pp 88–2646

Fieberg C, Kneer R (2008) Determination of thermal contact resistance from transient temperature measurements. Int J Heat Mass Transfer 51:1017–1023

Fletcher LS, Peterson GP, Madhusudana CV, Groll E (1989) Heat transfer through bolted and riveted joints. ASME HTD 123:107–115

Fletcher LS, Smuda PA, Gyorog DA (1969) Thermal contact resistance of selected low-conductance interstitial materials. AIAA J 7:1302–1309

Griesinger A, K Spindler K, Hahne E (1998) Measurements of the thermal contact conductance between a solid surface and a zeolite powder with the modified laser-flash method, heat transfer 1998. In: Proceedings of the 11th IHTC, vol. 4. p 101–106

Hohenauer W (1999) Laser flash method to determine thermal conductivity. http://phox.at

Howard JR (1976) An experimental study of heat transfer through periodically contacting surfaces. Int J Heat Mass Transfer 19:367–372

Hsu TR, Tam WK (1979) On thermal contact resistance of compound cylinders. In: AIAA 14th Thermophysics Conference. American Institute of Aeronautics and Astronautics, New York, pp 79–1069

Jeng DR (1967) Thermal contact resistance in vacuum. Trans ASME J Heat Transfer 89:275–276

Karplus WJ (1958) Analog simulation (solution of field problems). McGraw-Hill, New York, pp 140–170

Kempers R, Kolodner P, Lyons A, Robinson AJ (2009) A high precision apparatus for the characterization of thermal interface materials. Rev Sci Instrum 80:095111 (11 pages)

Madhusudana CV, Litvak A (1990) Thermal contact conductance of composite cylinders: an experimental study. J Thermophys Heat Transfer 4:79–85

Madhusudana CV (1992) The use of electrolytic tank for solving some problems in heat conduction. J Energy Heat Mass Transfer 14:59–64

Madhusudana CV (2000) Accuracy in Thermal Contact Conductance Experiments -The Effect of Heat Losses to the Surroundings, Int Comm Heat and Mass Trans 27:877–891

Madhusudana CV, Garimella SV (2003) Contact heat transfer measurements – steady state or transient?, Paper No.C2-179, ASME/JSME Thermal Engineering Joint Conference, Hawaii

Major SJ, Williams A (1977) The solution of a steady conduction heat transfer problem using an electrolytic tank analogue. Inst Eng (Australia) Mech Eng Trans 7–11

Mikic BB, Rohsenow WM (1966) Thermal contact resistance, Mech Eng Depart Report No. DSR 74542–41. MIT, Cambridge, MA

Milosevic' ND, Raynauld M, Maglic KD (2002) Estimation of thermal contact resistance between the materials of double-layer sample using the laser flash method. Inverse Prob Eng 10:85–103

O'Callaghan PW, Probert SD (1972) An improved thermal contact resistance rig. Meas Contr 5(5):311–315

Phillips GW Jr (1956) Effect of included media on thermal conductance of contact joints, MS thesis. US Naval Postgraduate School, Monterey, California, pp 17–18 and 34–37

Schneider GE, Strong AB, Yovanovich MM (1977) Transient thermal response of two bodies communicating through a small circular contact area. Int J Heat Mass Transfer 20:301–308

Sheffield JW, Sauer HJ Jr, Wood RA (1987) An experimental method for measuring the thermal contact resistance of plate finned tube heat exchangers. ASHRAE Trans 93(2):776–785

Veziroglu TN, Chandra S (1969) Thermal conductance of two dimensional constrictions. Prog Astronaut Aeronaut 21:591–615

Villanueva EP (1997) Thermal contact conductance, PhD thesis. The University of New South Wales

Williams A (1966) Heat transfer through metal to metal joints. In: 3rd International Heat Transfer Conference. American Institute of Chemical Engineers, New York, pp 109–117

Williams A, Madhusudana CV (1970) Heat flow across cylindrical metallic joints. In: 4th International Heat Transfer Conference, Paper Cu 3.6

Yip FC, Venart JES (1968) Surface topography effects in the estimation of thermal and electrical contact resistance. Proc I Mech Eng 182(3):81–91

Chapter 6
Special Configurations and Processes

In this chapter we will consider configurations other than simple plane joints and unique characteristics associated with them. In particular we will discuss in some detail contact conductance in bolted joints and cylindrical joints. This chapter also includes brief discussions on the role of thermal contact conductance in periodic contacts and in sliding friction.

6.1 Bolted or Riveted Joints

Bolted and riveted joints represent some of the most commonly used connections in engineering practice. It has been well known for a long time that when two plates are joined together by a central bolt, the contact area is limited to a relatively small annulus around the bolt hole (Rötscher 1927). This area will be called the "contact zone" in the following discussion. As will be described in the next section, recent theoretical and experimental analyses have indeed confirmed that the contact pressure decreases from a maximum near the edge of the bolt hole to nearly zero within a short radial distance (see Fig. 6.1). In Fig. 6.1, c and α define a conical envelope in which most of the variation in stress takes place and, therefore, α is sometimes called the cone dispersion angle. It is clear, therefore, that the total resistance to the axial heat flow through a bolted joint must consist of two parts:

a. A "macroscopic" resistance resulting from the constriction and spreading of the heat flow lines through the contact zone (Fig. 6.2), and
b. A "microscopic" resistance associated with the individual contact spots located within the contact zone.

It would thus appear that the extent of the contact zone, that is, the outer radius of the annulus should first be estimated before a thermal analysis is undertaken to determine the overall thermal resistance of a bolted or riveted joint. In other words, the stress distribution at the interface of the joint must first be determined.

C. V. Madhusudana, *Thermal Contact Conductance*,
Mechanical Engineering Series, DOI: 10.1007/978-3-319-01276-6_6,

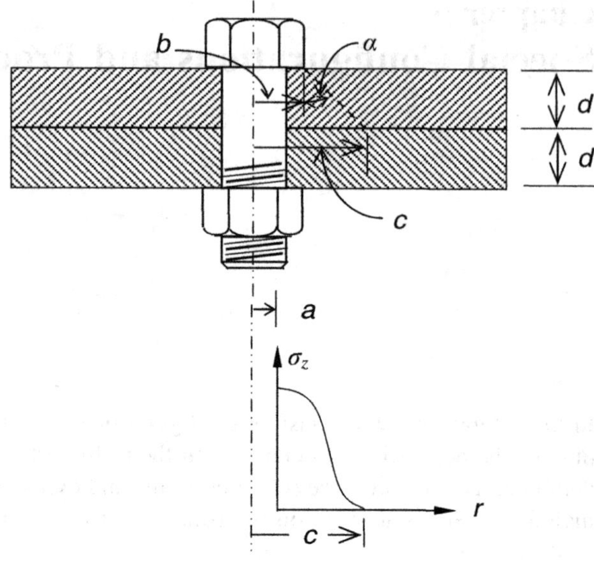

Fig. 6.1 Interfacial stress in a bolted joint

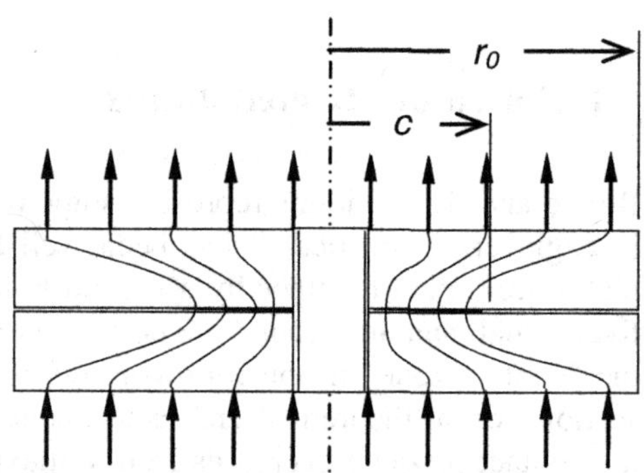

Fig. 6.2 Heat flow through a bolted joint

6.1.1 Stress Distribution at the Bolted Joint Interface

As an introduction to this problem, the simpler problem of a solid plate subjected to a uniform pressure over a central circular area will be considered first. The actual problem of two plates with central holes subjected to pressure over an annular area will be considered next.

One of the first mathematical investigations related to this problem was that described by Sneddon (1946). In this analysis, the stress distribution at different planes of a semi-infinite elastic medium subjected to a uniform pressure on parts of the boundary was considered. The solution to the biharmonic differential equation in cylindrical coordinates was obtained using Hankel transforms. The axial stress distribution at mid-plane ($z = 0$) for uniform pressure, p, applied over a circular area of radius, a, is shown in Fig. 6.3.

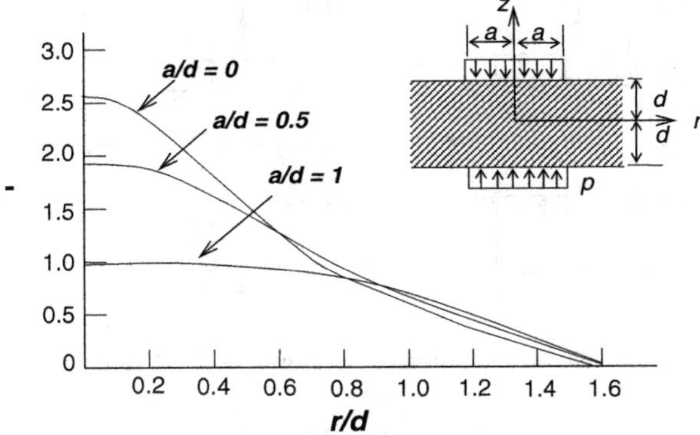

Fig. 6.3 Axial stress distribution at the mid-plane of a thick plate (Sneddon 1946)

Sneddon also proved that the axial stress, σ_z, was independent of the Poisson's ratio, ν, although the radial and circumferential stresses were slightly affected by the value of ν. These latter stresses, however, are not of significance to the present problem.

It was also shown that the result for the mid-plane of the thick plate could be used to determine the interface pressure distribution between the two plates of a bolted joint, provided the two plates were of equal thickness and made of the same material. Greenwood (1964) discussed the accuracy of Sneddon's solution and pointed out that they were accurate to values of r/d up to 0.6. In particular, he presented a table (see Table 6.1) for obtaining a good approximation of the radius of the contact circle: Thus, for $(a/d) > 0.5$, it is seen that

$$c \cong a + d \tag{6.1}$$

Lardner (1965) calculated the stresses as well as vertical displacements in a thick plate with axisymmetric loading and confirmed the result that two plates, of equal thickness, pushed together by forces opposite to each other, will separate when the vertical stress at the mid-plane of a single plate of twice the thickness becomes tensile. Thus the results for a single plate are applicable to two plates of equal thickness up to the radius at which the stress becomes positive.

In a bolted joint, the pressure would be applied over an annulus, rather than over a central circular area, on plates with a central hole. The solution to the problem of a single plate with a central hole subjected to a pressure applied over an annulus can be obtained by the method of superposition (Fernlund 1961) (see Fig. 6.4).

Table 6.1 Approximation of the radius of a contact circle

a/d	0	0.5	1.0	1.5	2.0	2.5	3.0
c/d	1.566	1.693	2.028	2.471	2.949	3.438	3.933

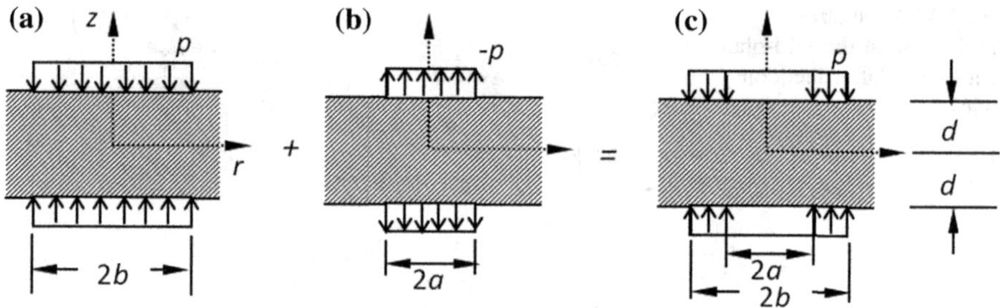

Fig. 6.4 The method of superposition (Fernlund 1961)

Note that, in Fig. 6.4c, although the pressure is applied over an annulus, the plate is still solid. To obtain the solution for the plate with a central hole (radius a), let

$$\begin{aligned}\sigma_r &= f_1(z)\\ \tau_{rz} &= f_2(z)\end{aligned} \tag{6.2}$$

at $r = a$. By further superimposing the normal stress, $-f_1(z)$ and the shearing stress, $-f_2(z)$ at $r = a$, we get a cylindrical surface of radius, a, that is stress free (Fig. 6.5). Thus the solution to the original problem is obtained as

$$(\sigma_z)_{\text{res}} = \sigma_z + \sigma_z^* + \sigma_{z2} + \sigma_{z3} \tag{6.3}$$

in which

$(\sigma_z)_{\text{res}}$ is the resultant axial stress (contact pressure)
σ_z is due to pressure, p, applied over $0 < r < b$
σ_z* is due to the tensile stress, p, applied over $0 < r < a$
σ_{z2} is due to the shearing stress—τ_{rz} applied at $r = a$
σ_{z3} is due to the normal stress—σ_r applied at $r = a$

Bradley et al. (1971) noted that the above superposition technique requires additional normal stresses at the surface of the plates and, consequently, the plates no longer have uniform annular loading. Thus, Fernlund's solution is approximate. Bradley et al. determined the interface stress distribution and contact area by a

Fig. 6.5 Fernlund's model of single plate with zero stress at $r = a$

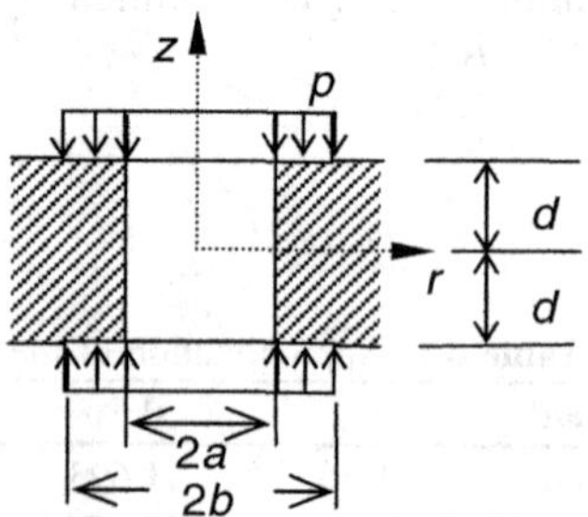

three-dimensional photoelastic analysis using the stress freezing technique. Results were presented for smooth flat plates of equal thickness. The results for $(a/d) = 0.5$ and $(b/a) = 1.5$ showed good agreement with results for the mid-plane stress in a thick plate obtained by a finite element analysis. It was noted that Fernlund's solution overestimated the contact radius and underestimated the maximum interface pressure by some 20 % (Fig. 6.6).

The theoretical analyses discussed so far have assumed that the bolted joint may be adequately modeled using the single plate model. The finite element and experimental analysis of Gould and Mikic (1972), however, showed that the single plate model yielded a contact zone that is larger than that obtained using a two-plate model, as shown in Table 6.2.

A series of papers dealing with the analytical solution of the problem of the thick plate with circular hole and axisymmetric loading was presented by Chandrashekhara and Muthanna (1977a, b; 1978; 1979). The solutions satisfied the exact boundary conditions rather than the approximate ones implied in Fernlund's superposition approach. The analytical results were expressed in terms of Fourier–Bessel series and integrals. Numerical computations showed that the results agreed well with those of Gould and Mikic for the single plate. Chandrashekhara and Muthanna (1978) also showed that the simpler two-dimensional solution for the semi-infinite strip gave results that were in substantial agreement with that for the more complex axisymmetric, three-dimensional problem for values of $(d/a) \leq 1$.

Motosh (1976) presented an approximate method for calculating the stress distribution, assuming that the bolt load dispersed in a conical or spheroidal envelope. The stress distribution was represented as a 4th degree polynomial (Eq. 6.4), in which the constants were determined from the boundary and equilibrium conditions. The "boundary conditions" were based on the assumptions made by Fernlund and previous experimental results from photoelastic models. However, Chandrashekhara and Muthanna (1978) noted the solution thus obtained agreed reasonably with the exact analytical solution *only at the mid-plane*, and

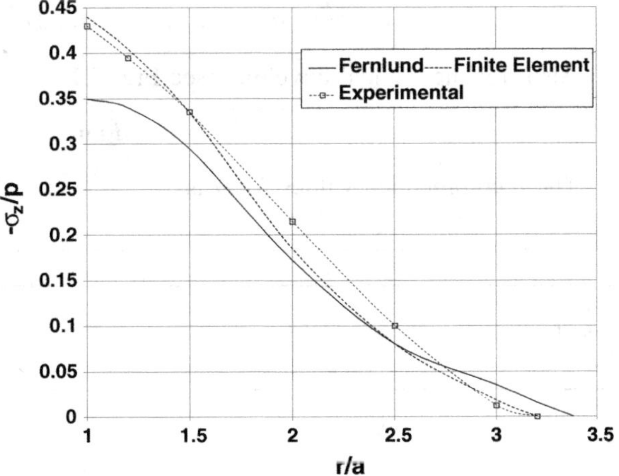

Fig. 6.6 Comparison of solutions for the stress at the mid-plane of a single plate (after Bradley et al. 1971)

Table 6.2 Contact zone: Single-plate model versus two-plate model

a/d	b/a	c/a		% discrepancy between models
		Single-plate model	Two-plate model	
1	3.1	4.2	3.7	13.5
	2.2	3.3	2.7	22.2
	1.6	2.7	2.1	28.6
	1.3	2.4	1.7	41.7
0.75	3.1	4.5	3.8	18.5
	2.2	3.6	2.8	28.9
	1.6	3.0	2.2	36.4
	1.3	2.7	2.0	35.0
0.5	3.1	5.1	4.1	24.4
	2.2	4.2	3.2	31.3
	1.6	3.6	2.8	28.6
	1.3	3.3	2.5	32.0

that, too, only for the conical envelope. Hence the conditions given below refer only to the mid-plane with load dispersed in a conical envelope. Inasmuch as we are interested only in the interface pressure distribution for the purpose of calculating the joint heat transfer, these results should be still useful in the present context.

$$\sigma = Ar^4 + Br^3 + Cr^2 + Dr + E \tag{6.4}$$

Equilibrium:

$$\int 2\pi\sigma r\,dr = \text{total joint load}$$

Boundary Conditions:

$$r = a: \quad \partial\sigma/\partial r = 0$$

$$r = c: \quad \sigma = 0, \partial\sigma/\partial r = 0 \text{ and } \partial^2\sigma/\partial r^2 = 0$$

Also, for the conical envelope (see Fig. 7.1),

$$c = b + d \tan \alpha$$

The recommended values for α are:

d/a	α (deg)
<2	≤40
2 to 4	45
≥4	≥50

6.1.2 Effect of Other Parameters

All of the above theoretical analyses have assumed that the interface is perfectly smooth. When the plate surfaces are rough, the width of the gap beyond the contact zone may be smaller than the surface asperities of either surface. This will cause compression of the interfering asperities, leading to an increase in the contact zone (Roca and Mikic 1972). This, in turn, will result in a change in the interface pressure distribution as illustrated in Fig. 6.7. These facts were confirmed by Ito et al. (1979), who experimentally determined the pressure distributions for lapped and ground surfaces by means of ultrasonic waves.

Minakuchi et al. (1983) supported their theoretical work on bolted joints (two plate model) by an experimental technique in which the interface pressure distribution was measured by means of pressure-sensitive pins. The effects of using different materials and also different thicknesses of plates were investigated. The results for the cone dispersion angles are presented in Table 6.3 (in all cases b/a was equal to 1.5):

From an inspection of Table 6.3, it is evident that:

1. The relative thicknesses of the plates can have a pronounced effect on the extent of the contact zone, and
2. The contact zone is not significantly affected by the use of different materials unless the two moduli of elasticity are vastly different.

The second of the above conclusions is important when soft gaskets are used in the bolted joint. Minakuchi (1984), in fact, confirmed that the contact area is much larger and the stress distribution "more gentle" when a soft metallic gasket is used.

The majority of the theoretical analyses assume that the loading, p, is uniform. Curti et al. (1985), using the boundary element method, determined the stress distributions for different loading (uniform as well as linearly varying with respect to the radius) conditions on plates of equal thickness. Although the interface pressure distribution was found to depend on the loading profile, it was noted that any such effect disappeared for $(b - a)/d < 0.3$. In other words, the exact nature of the applied pressure distribution is immaterial for "thick" plates. This merely confirms the earlier observation made by Fernlund (1961) on the basis of St. Venant's principle.

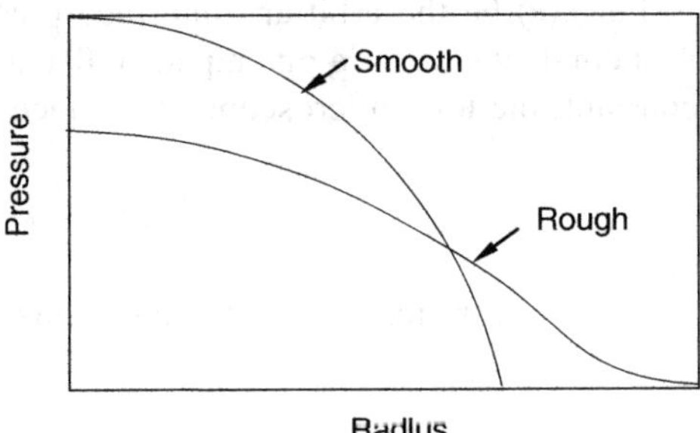

Fig. 6.7 Effect of surface roughness on contact zone

Table 6.3 Summary of cone dispersion angle results

$E_2/E_1 = 1$; $d_1/a = 3$			$d_2/d_1 = 1.5$; $d_1/a = 3$			$E_2/E_1 = 1$; $d_2/d_1 = 1$		
d_2/d_1	α_1 (deg)	α_2 (deg)	E_2/E_1	α_1 (deg)	α_2 (deg)	d_1/a	α_1 (deg)	α_2 (deg)
0.3	28	61	1.0	59	48	1	47	47
0.6	43	57	1.45	58	46	2	51	51
1.0	53	53	3.0	56	44	3	53	53
1.5	59	48	6.0	54	43	4.5	59	59
3.0	71	45	12.0	54	42	6	55	55

In their experimental study of bolted joint heat transfer, Oehler et al. (1979) noted that the axial tension in the bolt decreased with torquing cycles. Typically, about 10 cycles needed to be applied before a constant preload was reached. Oehler et al. also noticed that the interface pressure slightly increased at elevated temperatures due to the differential expansion between the bolt and the plates. Recent experimental and numerical analyses of Kumano et al. (1994) have shown that such increases could be very significant. Even for moderate temperature rises of the order of 100 to 200 °C, the corresponding increment in bolt tension could be greater than the initial preload.

6.1.3 Heat Transfer in Bolted Joints

As indicated at the beginning of this section, for axial heat flow through a bolted joint the total joint resistance is given by:

$$R_{\text{tot}} = R_{\text{micro}} + R_{\text{macro}} \tag{6.5}$$

Considering first the microscopic resistance, at first glance it would appear that this resistance must depend upon the interface pressure distribution within the contact zone. Yip (1972), indeed, considered three different stress distributions; namely, uniform, linear, and parabolic. Using the theoretical model for the contact conductance for nominally flat surfaces, he found that the three distributions yielded virtually the same resistance for any given load. Madhusudana et al. (1990a, b) proved why this should be so as follows:

Let $p(r)$ be the arbitrary interface pressure distribution. Then, since the solid spot conductance for a given pair of flat surfaces is given by $h_s = kp^n$, where k is a constant, the total microscopic resistance for the joint is given by

$$h_{\text{tot}} = \int kp^n (2\pi r\, dr)$$

In the theoretical model for flat joints, n is nearly equal to 1 and, hence

$$h_{tot} = k \int p(2\pi r\, dr)$$

But

$$\int p(2\pi r\, dr) = \text{total mechanical load on the joint}$$

The total microscopic conductance or resistance for a given load must, therefore, remain the same irrespective of the pressure distribution.

Madhusudana et al. further pointed out that, if *experimental correlations* for the solid spot conductance are used, then, because of the flatness deviations and other surface irregularities inevitably present in practical surfaces, the value of n is smaller than unity (typically 0.6 to 0.7). In such a case, there may be a noticeable difference between the conductance values obtained using different stress distributions (Fig. 6.8). Again, this may not be significant for most applications. Referring to the same figure, it is important to note, however, that the microscopic conductance increases with the extent of the contact zone (defined by the radius, c). Inasmuch as the macroscopic conductance also increases with the radius, c, it is apparent that this radius is the single most important parameter defining the thermal conductance of a bolted joint for a given pair of surfaces. As seen during the discussion of the interface pressure distribution in the bolted joint, this radius itself is found to depend on the bolt hole radius, b, and the thickness, d, of the plates by means of an equation of the form:

$$c = b + (\text{constant})\, d$$

It is thus clear that the bolt hole radius and the plate thickness are two important parameters to be considered in the thermal design of a joint Fig. 6.8.

It may also be recalled that the presence of surface roughness tended to increase the radius, c. Although this would increase both the microscopic and macroscopic conductances, the roughness would also tend to increase the microscopic resistance. Hence Roca and Mikic (1972) showed that it is possible to control the overall conductance by a suitable choice of roughness in relation to the other system parameters.

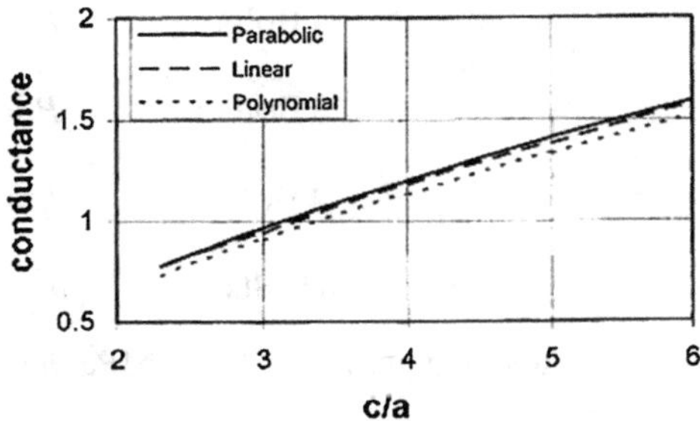

Fig. 6.8 Effect of pressure distribution on microscopic conductance

Once the contact zone radius has been determined, it is a straightforward matter to determine the macroscopic resistance associated with the contact zone. The macroscopic resistance of a bolted joint in vacuum, for example, may be simulated by a simple electrolytic tank analogue (Fletcher et al. 1990). The macroscopic resistance in a vacuum or in a conducting medium may be determined using a finite difference technique (Madhusudana 1994). Figures 6.9a, b ,c shows typical results obtained from the finite difference analysis. In this figure, x is the ratio of the conductivity of the fluid in the gap ($c < a < r_o$) to the conductivity of the material of the plates. The macroscopic resistance is nondimensionalized as:

$$R_N = (R - R_S)/R_S \tag{6.6}$$

in which

R = total resistance with the macroscopic constriction
R_S = total resistance without the constriction; i.e., the resistance of a hollow cylinder of inner radius, c, outer radius, r_o, and thickness equal to the sum of the thicknesses of the two plates.

It is clear that the macroscopic resistance decreases with increase in both the thickness and the contact zone radius. In an actual joint, however, the thickness and the contact zone radius are not independent, as mentioned before. It is also seen that for $x \le .0001$, the resistance is virtually the same as for vacuum ($x = 0$). The results in Fig. 6.9c for vacuum conditions show good agreement with the experimental results of Fletcher et al. (1990).

The experimental work of Yeh et al. (2001) has already been referred to in Chap. 1, Introduction. The experimental details were as follows:

Aluminium alloy (6061-T6) Blocks: 63.5 mm × 63.5 mm, height 50 mm fastened together by aluminium bolts with hexagonal nuts.

Typical range of surface roughness, σ = 0.25–0.4 μm

Bolt diameters: 3, 5 and 8 mm

Torque Range: 1 to 10 Nm.

Their results confirmed that the contact pressure across a bolted joint is not uniform—the pressure is large near the bolt (holes). They also found that the contact pressure and, therefore, the TCC, increased with (a) the bolt torque and (b) the number of bolts as indicated in Figs 6.10 and 6.11.

The derived correlations are also listed below:

$$4\,\text{Bolts}: \qquad P = 1.40 + 1.02\tau^{0.28} \tag{6.7}$$

$$8\,\text{Bolts}: \qquad P = 1.94 + 0.70\tau^{0.59} \tag{6.8}$$

$$4\,\text{Bolts}: \qquad h = 3.89 + 0.29\tau^{1.24} \tag{6.9}$$

$$8\,\text{Bolts}: \qquad h = 12.67 + 0.45\tau^{2.01} \tag{6.10}$$

In the above equations, P is in MPa, h is in kW/(m^2 K) and the torque τ is in Nm.

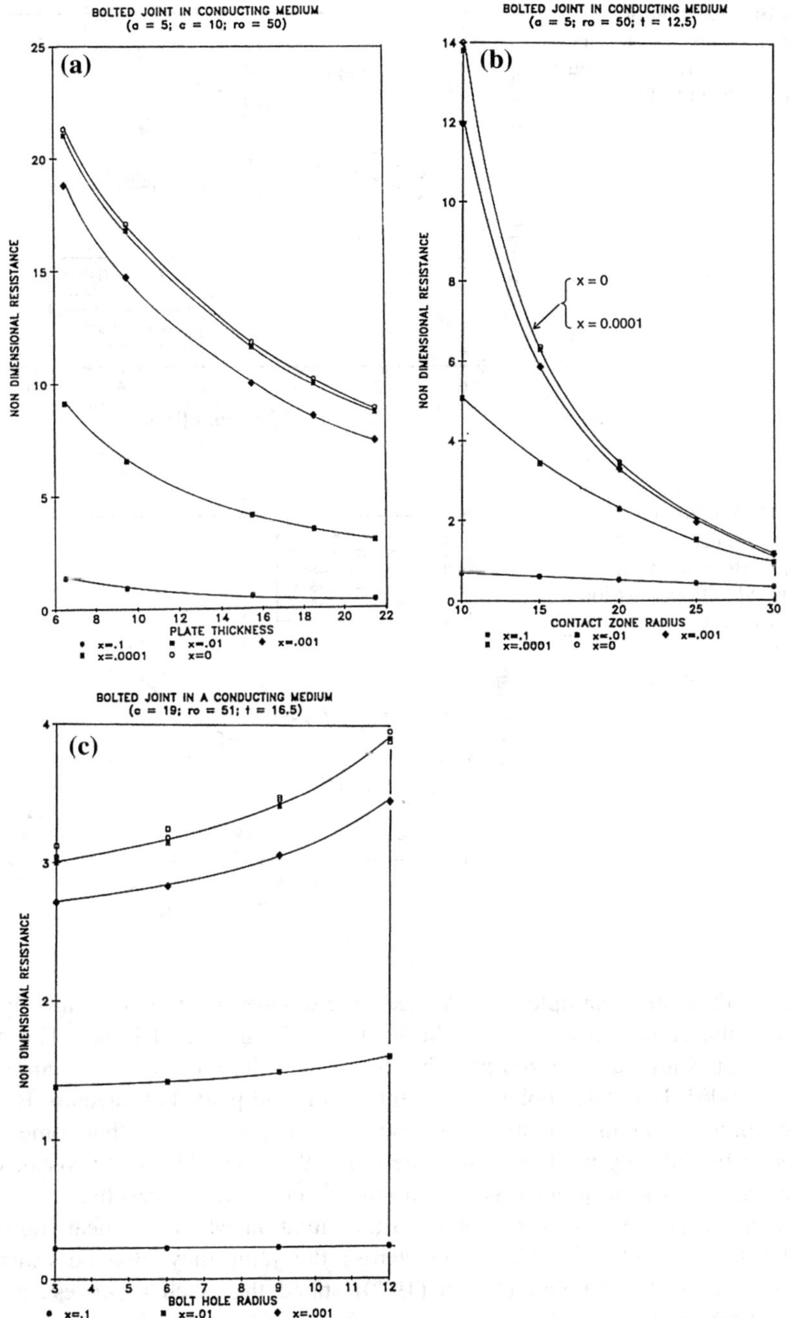

Fig. 6.9 **a** Nondimensional resistance versus plate thickness, **b** Nondimensional resistance versus contact zone radius, **c** Nondimensional resistance versus bolt hole radius; [Experimental results for vacuum from Fletcher et al. (1990)]

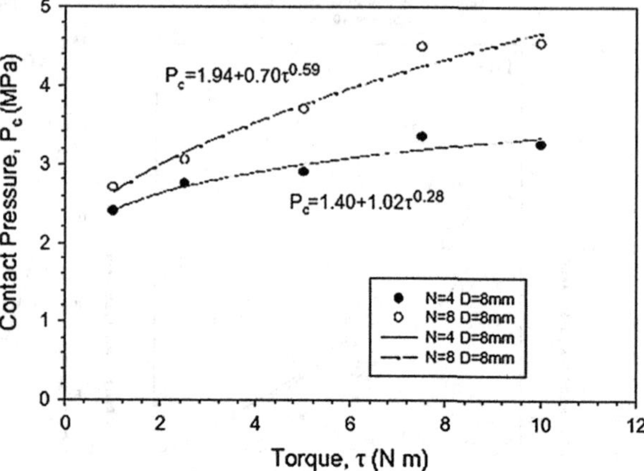

Fig. 6.10 Variation of contact pressure with torque (Yeh et al. 2001, reproduced with permission from Elsevier)

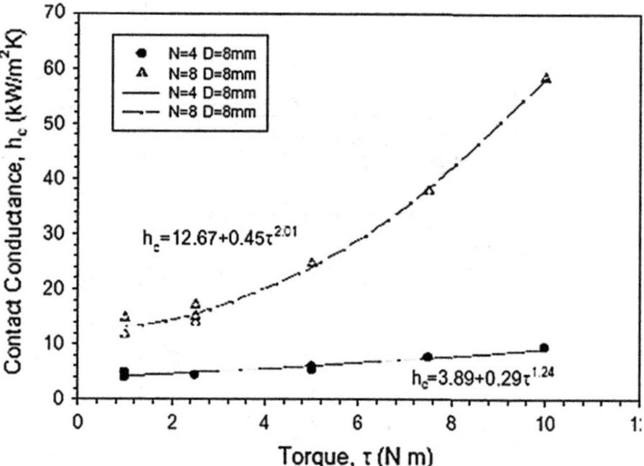

Fig. 6.11 Variation of contact conductance with torque (Yeh et al. 2001, reproduced with permission from Elsevier)

Other than the example just quoted, there have been comparatively few experimental studies dealing with the heat transfer in bolted joints. The representative ones are summarized in Table 6.4. All of these tests refer to aluminum alloy (A16061-T6) plates bolted by a central bolt and placed in vacuum. Because of the differing loading conditions, surface texture, and the fact that some investigators reported only the total resistance rather than the additional resistance due to the presence of the joint, it is not possible to compare the results.

All of the above discussion refers to the situation where the heat transfer is parallel to the bolt axis. Heat flow across the joint may also be transverse (Fig. 6.12). Whitehurst and Durbin (1970) stated that such a process may be analysed by defining an effective thermal conductivity for the joint:

$$k_e = \frac{X_2}{X_e} k \tag{6.11}$$

Table 6.4 Heat transfer in bolted joints; Parameters used in some representative studies

Parameter	Aron and Colombo (1963)	Elliott (1965)	Veilleux and Mark (1969)	Oehler et al. (1979)	Mittlebach et al. 1993
Plate thickness, d, mm	1.5	2.1	top: 1 bottom: 1.25	top: 29 bottom: 35.5	top: 25.4 bottom: 19.1
Bolt diameter, $2a$, mm	4.8	4.8	3.5	6.35	3.175
Loading radius, b, mm	5.5	5.5	3.4	≅6.35	4.77
Roughness (rms), μm	0.2 to 0.55	0.5 to 0.75	0.4 to 0.5	0.8	top: 0.64 bottom: 0.71
Bolt load, N	450 to 5800	900 to 5800	1600	6700	2000 to 6400 kPa

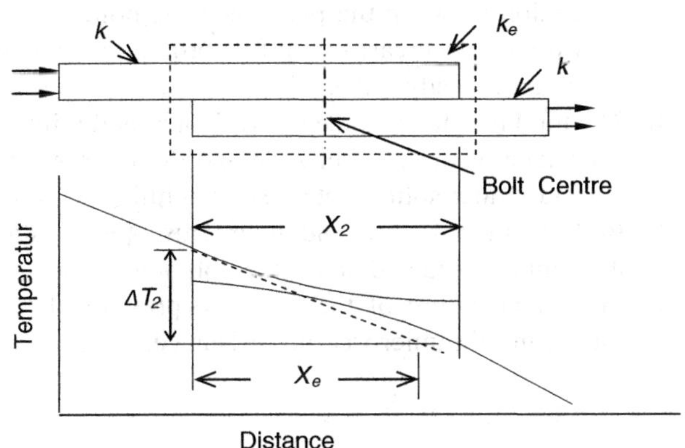

Fig. 6.12 Transverse heat flow in a bolted joint

This method assumes one-dimensional heat flow without losses and requires temperature gradients be determined (experimentally) for the specific joints under consideration or, alternatively, a large number of models constructed and tested. Roca and Mikic (1972) presented the results of a numerical solution of a similar problem. Lee et al. (1993) described an analytical solution of the latter problem, assuming that the extent of the contact zone was known, and that perfect contact existed over this zone.

6.1.4 Summary

The following conclusions follow as a result of the discussion in this section.

1. The interface pressure distribution in a bolted joint is nonuniform, with the peak pressure near the edge of the bolt hole, rapidly decreasing to zero in a short radial distance.
2. The outer radius of the contact zone depends on the bolt hole radius and the plate thickness. It is independent of the axial bolt load.
3. The contact zone, as given by the two-plate analysis, is smaller than that obtained using a single-plate model.
4. The contact zone obtained for rough surfaces is larger than that obtained assuming the surfaces to be perfectly smooth.
5. For two plates made of different materials, the contact zone is not affected by the different moduli of elasticity, unless the two moduli are vastly different.
6. At elevated temperatures, the bolt load is increased due to differential expansion between the plates and the bolt.
7. Bolt preload decreases with the torquing cycles for the first few cycles before reaching a steady value.
8. The total resistance to heat flow in a bolted joint is the sum of the macroscopic resistance of the contact zone and the microscopic resistance associated with the individual solid contact spots within the contact zone.
9. Both the macroscopic and the microscopic resistances depend on the extent of the contact zone and hence the bolt hole radius and the thickness of the plates.
10. The exact nature of the interface pressure distribution is not important for estimating the microscopic resistance.

6.2 Cylindrical Joints

There are many applications in which the heat flow is radial across concentric, compound cylinders. Examples include plug and ring assemblies, shrink-fit cylinders, finned tube heat exchangers, duplex and multiplex tubes used in solar thermal power plants and nuclear reactor fuel elements. The experimental and theoretical investigations in each of these categories usually deal with specific applications. For example, the works of Sheffield et al. (1984, 1985, 1987, 1989) deal with mechanically expanded copper tubes with aluminum fins.

Reviews of literature on the basic mechanism of heat flow through cylindrical joints presented by Madhusudana et al. (1990a, b) and Ayers et al. (1997) point out the fact that, although cylindrical joints are just as commonplace as flat joints, the available literature on heat flow through cylindrical joints is comparatively small. A main reason for this appears to be the additional complexities that are present in a cylindrical joint.

In heat transfer across flat joints, the contact pressure is usually known (or estimated) and can be directly controlled. It is, therefore, chosen both in theory and experiment, as the independent variable controlling the conductance of a given joint. In a cylindrical joint, however, the contact pressure and, therefore, the contact conductance depend on the interference between the cylinders existing at the time of operation (Williams and Madhusudana 1970).

This interference consists of the following components:

1 The differential expansion, u_A, due to the temperature gradients caused by the heat flow; u_A can be calculated using equations of thermoelasticity.
2 The differential expansion, u_B, caused by the fact that, because of the contact resistance, the two surfaces at the interface will be at different temperatures.
3 The interference, u_C, at the time of assembly (that is, the initial degree of fit).

It can be seen that u_A depends on the heat flux and u_B depends on both the heat flux and the operating temperatures. The heat flux, or some measure of it, and the maximum temperatures are, therefore, the primary variables controlling the contact conductance of a given pair of cylinders. Thus, the heat flux and the maximum temperature, rather than the indirectly estimated contact pressure, become the logical independent variables in the analysis of cylindrical joint thermal conductance.

In many subsequent works, the heat flux has, indeed, been used as the independent variable (Hsu and Tam 1979; Egorov et al. 1989).

The following theory (Madhusudana 1999) applies to radially outward flow and makes use of the following simplifying assumptions:

1. The surfaces are rough but conforming; there are no large scale irregularities such as out-of-roundness or waviness.
2. The heat transfer rate across the gas gaps is constant; in other words, the variation of the effective thickness of the gap with heat flux is considered negligible.
3. Heat transfer by radiation is ignored.

Referring to Fig. 6.13, the pressure developed between the two cylinders as a result of the interference between them is given by (Timoshenko 1956):

$$\left(\frac{u}{b}\right) = \frac{P}{E_i}\left\{\frac{E_i}{E_0}\left[\frac{c^2+b^2}{c^2-b^2}+\nu_o\right]+\left[\frac{b^2+a^2}{b^2-a^2}-\nu_i\right]\right\} \tag{6.12}$$

or

$$\left(\frac{u}{b}\right) = P\{C_1\} \tag{6.13}$$

in which

$$u = u_A + u_B + u_C \tag{6.14}$$

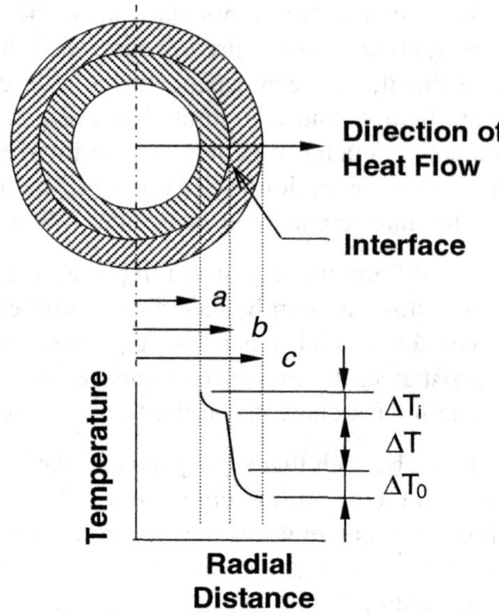

Fig. 6.13 Heat flow through a cylindrical joint

and

$$C_1 = \frac{1}{E_i}\left\{\frac{E_i}{E_0}\left[\frac{c^2+b^2}{c^2-b^2}+v_o\right]+\left[\frac{b^2+a^2}{b^2-a^2}-v_i\right]\right\} \tag{6.15}$$

For a given pair of cylinders the initial interference, u_C, is known. The differential expansions u_A and u_B are to be determined from the conditions of heat flow.

Interference Due to Heat Flow

The steady state heat flow is:

$$Q = \frac{2\pi L k_i \Delta T_i}{ln(b/a)} = \frac{2\pi L k_o \Delta T_o}{ln(c/b)} \tag{6.16}$$

where L is the axial length of the composite cylinder.

Hence

$$\Delta T_o = \left(\frac{k_i}{k_o}\right)\left[\frac{ln(c/b)}{ln(b/a)}\right]\Delta T_i \tag{6.17}$$

The heat flow results in the deformations of the cylinders. At the interface, the radial displacements of the inner and outer cylinders are, respectively (Timoshenko and Goodier 1970):

$$u_{Ai} = \frac{b\alpha_i \Delta T_i}{2ln\left(\frac{b}{a}\right)}\left[1 - \frac{2a^2}{b^2 - a^2} ln\left(\frac{b}{a}\right)\right] \tag{6.18}$$

$$u_{Ao} = \frac{b\alpha_o \Delta T_o}{2ln\left(\frac{c}{b}\right)}\left[1 - \frac{2c^2}{c^2 - b^2} ln\left(\frac{c}{b}\right)\right] \tag{6.19}$$

Substituting for ΔT_o in the expression for u_{Ao} and simplifying

$$u_{Ao} = \frac{b\alpha_i \Delta T_i}{2\ln\left(\frac{b}{a}\right)}\left(\frac{\alpha_o k_i}{\alpha_i k_o}\right)\left[1 - \frac{2c^2}{c^2 - b^2} ln\left(\frac{c}{b}\right)\right] \tag{6.20}$$

Since the nett interference due to heat flow is

$$u_A = u_{Ai} - u_{Ao} \tag{6.21}$$

we can write

$$\frac{u_A}{b} = \Delta T_i(C_2) \tag{6.22}$$

where

$$C_2 = \frac{\alpha_i}{2ln\left(\frac{b}{a}\right)}\left\{\left[1 - \frac{2}{\frac{b^2}{a^2} - 1} ln\left(\frac{b}{a}\right)\right] - \left(\frac{\alpha_o k_i}{\alpha_i k_o}\right)\left[1 - \frac{2}{1 - \frac{b^2}{c^2}} ln\left(\frac{c}{b}\right)\right]\right\} \tag{6.23}$$

Interference due to contact resistance

Because of the finite resistance to the heat flow at the joint, the two sides of the interface will be at different temperatures. The consequent differential expansion is:

$$u_B = u_{Bi} - u_{Bo} == b\alpha_i T_1 - b\alpha_o (T_1 - \Delta T) \tag{6.24}$$

or

$$\frac{u_B}{b} = [T_1(\alpha_i - \alpha_o) + \alpha_o \Delta T] \tag{6.25}$$

Here T_1 is the temperature rise of the outer surface of the inner cylinder due to heat flow, and ΔT is the temperature drop due to contact resistance.

If T_a refers to the *temperature rise of the inner surface of the inner cylinder*, then

$$\frac{u_B}{b} = [(T_a - \Delta T_i)(\alpha_i - \alpha_o) + \alpha_o \Delta T] \tag{6.26}$$

Note that T_a is measured with respect to the temperature T_∞ at which the dimensions of the cylinders were established (e.g., the room temperature). In other words

$$T_a = T_{max} - T_\infty$$

where T_{max} is the maximum temperature obtained during heat flow.

Also note that if the two cylinders are made of the same material (or, at least, their coefficients of thermal expansion are the same), then the first term inside the brackets of Eq. (6.26) becomes zero. In this case, the maximum temperature has no significance (there will be, of course, some variation if the change in material properties and radiation at higher temperatures are taken into account) and u_B is specified by only ΔT, which is given by

$$\Delta T = \frac{q}{h} \tag{6.27}$$

where

$$q = \frac{\frac{2\pi L k_i \Delta T_i}{ln(b/a)}}{(2\pi L b)} = \frac{k_i \Delta T_i}{b ln(b/a)} \tag{6.28}$$

and

$$h = h_s + h_g \tag{6.29}$$

Whether we use Mikic's derivation for ideal flat surfaces or empirical correlations based on experimental results such as Tien's, the solid spot conductance can be written in the form;

$$h_s = C_3 \tan\theta \left(\frac{k}{\sigma}\right)\left(\frac{P}{H}\right)^n \tag{6.30}$$

The gas gap conductance may be calculated as

$$h_g = \frac{k_g}{\delta_{\text{eff}}} \tag{6.31}$$

The effective thickness, δ_{eff}, of the gas gap must take into account both the dimensions of the physical gap as well as the temperature distance. A reasonable approximation for δ_{eff} (see Sect. 4.6) is:

$$\delta_{\text{eff}} \approx 3\sigma \tag{6.32}$$

However, no numerical values need to be used during derivation.

Eq. (6.26) may therefore be written as

$$\frac{u_B}{b} = C_4(T_a - \Delta T_i) + \frac{\alpha_o k_i \Delta T_i}{b ln(b/a)} \left[\frac{1}{C_3 \tan\theta \left(\frac{k}{\sigma}\right)\left(\frac{P}{H}\right)^n + \frac{k_g}{\delta_{\text{eff}}}}\right] \tag{6.33}$$

or

$$\frac{u_B}{b} = C_4(T_a - \Delta T_i) + C_5\left[\frac{1}{C_6P^n + C_7}\right]\Delta T_i \tag{6.34}$$

where

$$C_4 = \alpha_i - \alpha_o$$

$$C_5 = \frac{\alpha_o k_i}{b ln(b/a)}$$

$$C_6 = C_3 \tan\theta\left(\frac{k}{\sigma}\right)\left(\frac{1}{H}\right)^n$$

$$C_7 = \frac{k_g}{\delta_{\text{eff}}}$$

Thus the governing equation for contact pressure can be written as

$$PC_1 = C_2\Delta T_i + C_4(T_a - \Delta T_i) + C_5\left[\frac{1}{C_6P^n + C_7}\right]\Delta T_i + \frac{u_C}{b} \tag{6.35}$$

As P appears on both sides of this equation, an iterative method is necessary to solve for P. However, since the value of n is close to 1, we can choose $n = 1$ as a first approximation. This will yield a simple quadratic in P which can be refined if necessary.

The following numerical example will illustrate the theory:

6.2.1 Numerical Example

Data:

1. *Material Properties*		
Material	Stainless steel	Aluminium alloy
Modulus of Elasticity, GPa	200	70
Poisson's ratio	0.28	0.33
Microhardness	3800	1400
Coefficiient of thermal expansion (1/˚C)	18×10^{-6}	24×10^{-6}
Thermal conductivity [W/(mK)]	16.5	200
Thermal conductivity of Air [W/(mK)]	0.0262	

2. *Geometric and Surface Properties*				
a (mm)	b (mm)	c (mm)	σ (μm)	tanθ
9	10	11	3	0.176

With the details as given the constants in Eq.(6.35) can be calculated. Results are shown in the table below:

	Stainless steel → Aluminium	Aluminium → Stainless steel
C1	0.000201	0.000185
C2	9.61×10^{-6}	113×10^{-6}
C3	0.55	0.550
C4	-6×10^{-6}	6×10^{-6}
C5	0.37585	3.4168
C6	2087	2087
C7	2911	2911

Note 1. In calculating C6, Tien's correlation is used; but results using Mikic's theory has also been calculated and shown in one of the graphs below for comparison.

2. Gas gap conductance, C7, has been approximated by hg ≈ (kg/3σ); but more sophisticated results could be used if necessary.

Some results are shown in the graphs below. Figure 6.14 illustrates the change in contact pressure and conductance with the heat flow rate when the heat flow direction is from stainless steel inner cylinder to aluminium outer cylinder. The maximum temperature in the system, that is, the temperature of the inner cylinder is T_a taken to be 100 °C, measured above the ambient, for this example. Tien's empirical correlation was used to calculate the solid spot conductance.

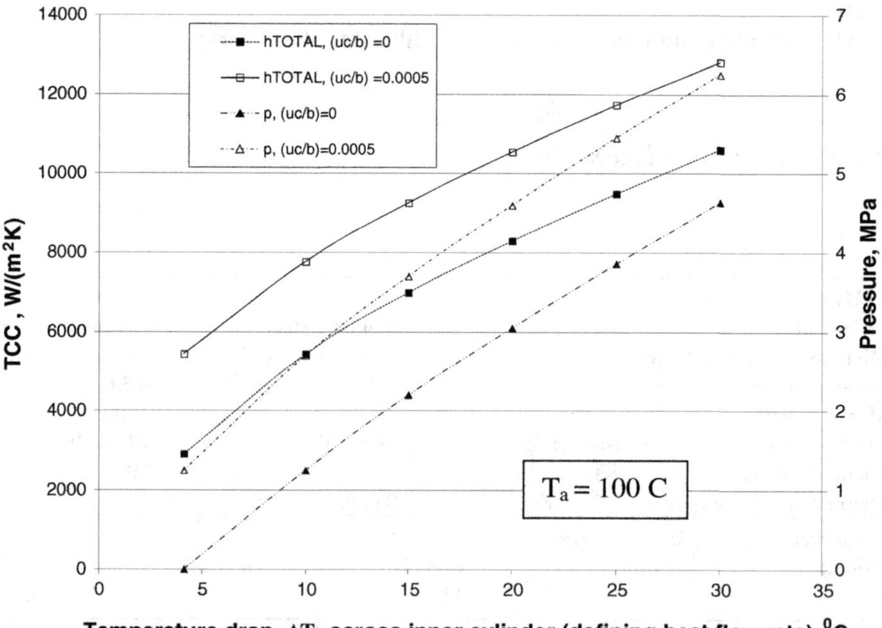

Fig. 6.14 Variation of contact pressure and conductance with heat flow rate; effect of changing the initial interference. Stainless → Aluminium in air

For the cylinders assembled with zero clearance, a minimum heat flow is required (in this case, equivalent to $\Delta T_i = 4.15$ °C, is required to close the gap (that is, the expansion due to heat flow overcomes the differences in expansions of the two cylinder). From thereon, the pressure is reinforced continuously. The conductance at zero contact pressure, of course, is the gap conductance due to the presence of air in the interstices. With an initial interference, a positive contact pressure already exists with zero heat flow, and the corresponding solid spot conductance needs to be added as shown.

Figure 6.15 shows the effect of varying the maximum temperature for the same configuration. The heat flow rate, equivalent to $\Delta T_i = 10$ °C, is kept constant and the initial clearance taken to be zero for this illustration. The effect of using different correlations (Tien's and Mikic's) for the calculation of solid spot conductance can be also visualized in this figure. It is noted that the difference is marginal. This is because of the interdependence of the contact pressure and the heat flux. In other words, for cylindrical joints, the exact form of the correlation used for the solid spot conductance is relatively unimportant, compared to that for flat joint. Note that, for a given heat flow rate, the contact pressure is progressively relaxed, and the solid spot conductance correspondingly reduced, as the maximum temperature is raised. It can be verified that the pressure becomes zero for $T_a \approx 241$ °C in this example (zero initial clearance), from which point the conduction will be by air in the gaps alone. (If there was an even a slight initial interference, for example, $u_C/b = 0.0005$, the temperature rise at which the pressure becomes zero increases to 324 °C).

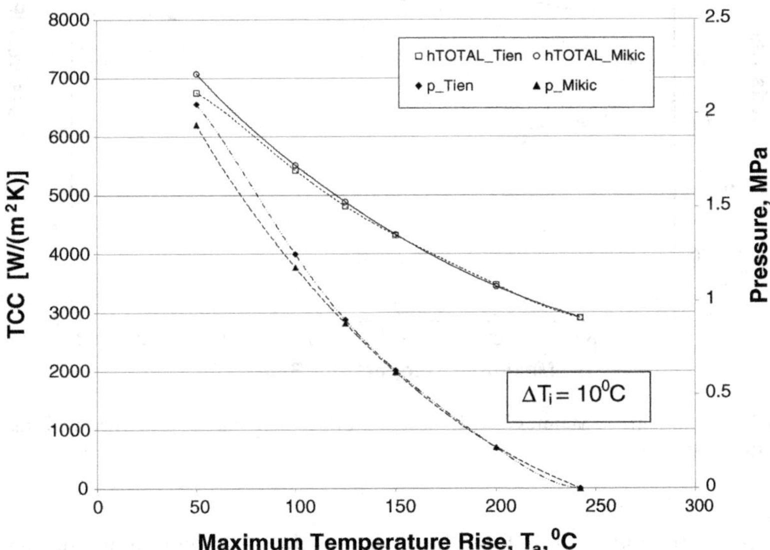

Fig. 6.15 Variation of pressure and conductance with maximum temperature (Stainless steel → Aluminium); effect of using different correlations for the solid spot conductance is also shown

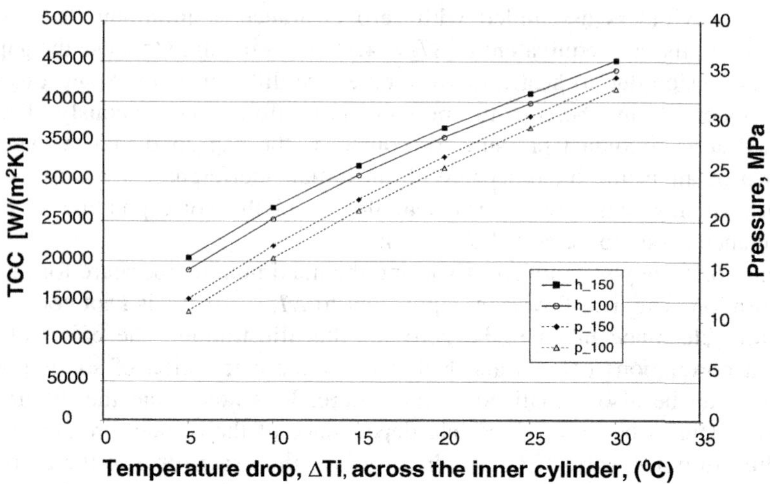

Fig. 6.16 Aluminium → Stainless steel in air; Variation of Pressure and Conductance with heat flow rate for Ta = 100 and Ta = 150

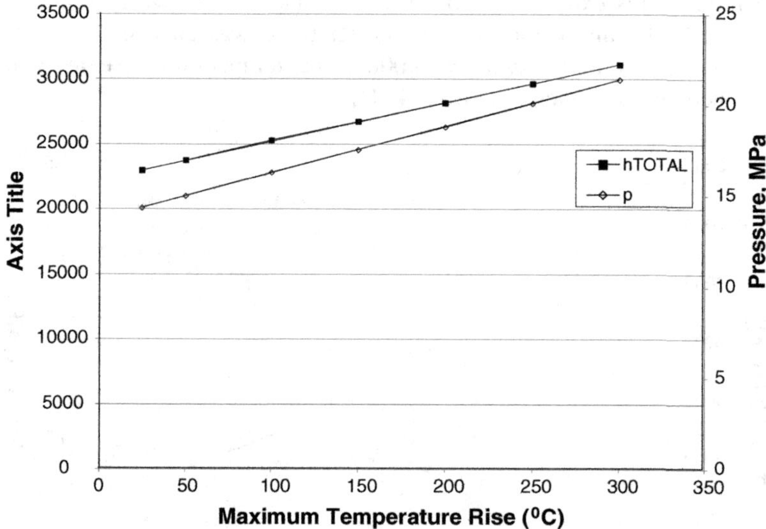

Fig. 6.17 Aluminium → Stainless steel; variation of pressure and conductance with maximum temperature rise

Figures 6.16 and 6.17 below show the results of calculations for the same material combination, but with the inner and outer cylinders exchanged. It can be seen that with aluminium inner cylinder and radially outward heat flow:

- the conductance and pressures achieved are much larger than those achieved with stainless steel inner cylinder
- the contact pressure is reinforced continuously both with the heat flow rate and the maximum temperature within the system—the contact pressure will never be reduced to zero.

For *thin* cylinders or tubes, both the thermal stress and the heat flux equations will be much simpler because linear approximations could be used for thin cylinders. This will be illustrated in discussing some experimental results in the section on finned tube heat exchangers.

6.3 Periodic Contacts

Periodic contact heat transfer occurs, for example, between the valve and seat in an internal combustion engine, a soldering iron and work piece on an assembly line, and a hot work-piece and die under repetitive forming conditions. The thermal contact resistance between two periodically contacting surfaces can be the most significant factor in controlling the heat flow, particularly when the ratio of the contact time to cycle time is high (Howard and Sutton 1973).

The characteristic feature of the quasi-steady heat transfer between periodically contacting solids is that the linear temperature distribution is retained in the bulk of solids except for some fluctuations below the contacting surfaces. The depth of the temperature fluctuation depends on frequency of contact and, for parts of internal conbustion engines, it is usually of the order of a fraction of a millimetre.

In a pioneering study, Howard and Sutton (1970) analyzed the problem of two bars in periodic contact with the help of an analogue computer. They made the following assumptions in their analysis:

- Heat flow was one dimensional.
- The bars were in perfect contact (no contact resistance) during the contact period and perfect separation during the non-contact period of the cycle.

These are illustrated in the Fig. 6.18 and boundary conditions below. The subscripts H and C refer to the hot and the cold bars, respectively. τ_c is the duration of contact per cycle and τ is the period (the sum of contact and non-contact times).

During contact period ($0 < t < \tau_c$): $T_{OH} = T_{OC}$; that is, no contact resistance.

During separation ($\tau_c < t < \tau$): $k_H(\partial T_H/\partial x) = k_C(\partial T_C/\partial x)$; that is, no heat transfer.

The governing equation is:

$$\frac{\partial T}{\partial t} = \alpha \frac{\partial^2 T}{\partial x^2}$$

Howard and Sutton further assumed that the bars *were made of same material* so that one side of the interface only needed to be analysed. They considered that the

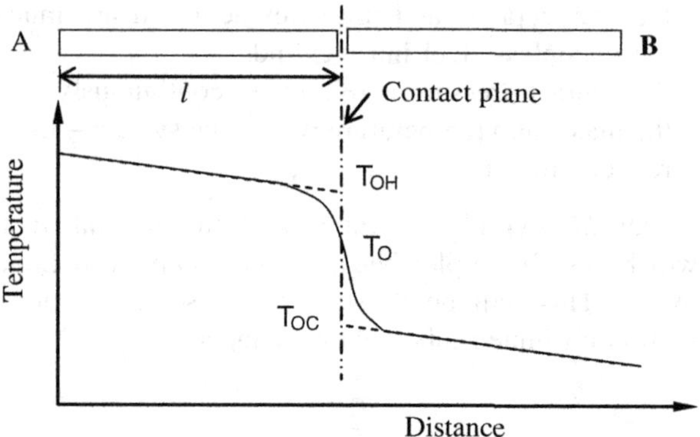

Fig. 6.18 Two bars in periodic contact

thermal resistance of the system consisted of two *independent* thermal resistances in series, R_s and R_i. R_s is the resistance under permanent contact conditions (due to the length, l, of the bar only, if the contact resistance is neglected) and R_i is the additional resistance due to periodic interruption to the heat flow. This additional length may be represented by an equivalent length l_i of the same material and same cross-sectional area (see Fig. 6.19).

The equivalent length l_i depends on the length of the bar l, the diffusivity α, the frequency of contact f and the duration of contact per cycle, τ_c. The relationship between the variables was expressed in the following non dimensional form:

$$\left(\frac{fl_i^2}{\alpha}\right) = g\left[\left(\frac{fl^2}{\alpha}\right), (f\tau_c)\right]$$

Once a quasi steady state has been reached, the temperature fluctuations are confined to a short distance from the interface. For sinusoidal variation, the amplitude of temperature at a distance δ (as given below) is less than 0.66 % of that at the contact plane (Eckert and Drake 1959).

$$\delta = 1.6\sqrt{\left(\frac{\pi\alpha}{f}\right)}$$

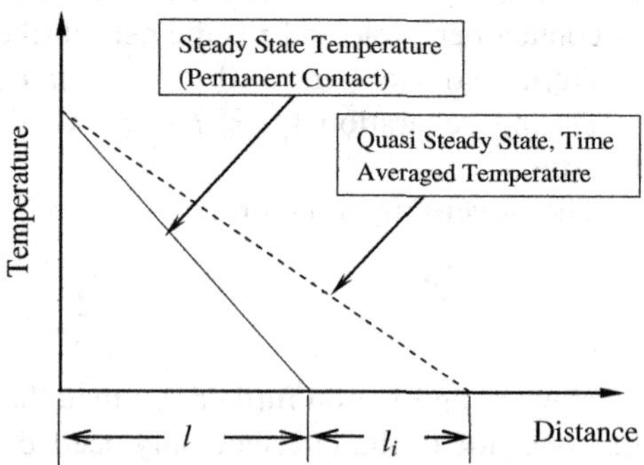

Fig. 6.19 Physical length of bar and the equivalent length to account for periodic disruption to heat flow

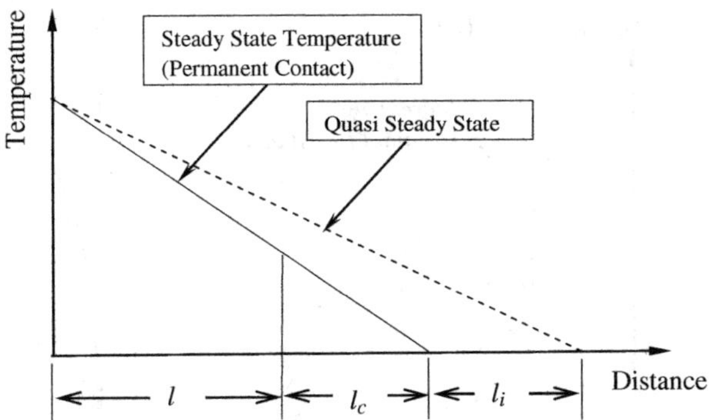

Fig. 6.20 Contact resistance added to permanent and periodic resistances

Hence it can be assumed that l_i is independent of l when $(fl^2/\alpha) > 2\pi$ and (fl_i^2/α) is a function of $(f\tau_c)$ only. Howard and Sutton solved the problem by the use of an analogue computer and finite difference method.

In a later work, Howard and Sutton (1973) refined their model to include the effect of contact resistance by adding another equivalent length, l_c to the physical length as shown in Fig. 6.20. The effect of contact resistance is to amplify the magnitude of the disruption to heat flow caused by periodic contact, as indicated in Fig. 6.21.

The experimental studies of Moses and Johnson (1988) focused on two basic areas: (a) the behaviour of thermal contact conductance during the quasi-steady state, and (b) the length of time required for the temperature distribution in the material to approach that observed in the quasi-steady condition. Their experimental apparatus has been described in Chap. 5.

Moses and Johnson used the following non-dimensional parameters for the presentation of their results:

Fourier Number, Fo $= (\alpha t/L^2)$, where L is the specimen length.

Relative contact conductance $= h/h_{ss}$, where h_{ss} is the steady state conductance

Non dimensional contact time parameter: $\tau^* = (\alpha\tau_c/L^2)$

Contact conductance parameter (Biot Number), Bi $= (hL/k)$

The results of Moses and Johnson for brass/brass contacts are indicated in Fig. 6.22. The graph represents, approximately, the combined results for four values (0.2–0.5) of the contact time parameter. For all quasi steady cases—each with a value of Fo < 1—the contact conductance was found to increase throughout the contact period (this actually results in the non-dimensional conductance temporarily overshooting the value of 1 before returning to a value less than 1). Hence the common practice of assuming the thermal contact conductance to be constant throughout the contact period is open to question, especially if the duration of contact is small.

The length of time, or more appropriately, the number of cycles required to approach quasi-steady state is expected to depend on both the contact conductance and the duration of contact. Hence the number of cycles was plotted against the product of the contact time parameter, τ^*, and the Biot number as shown in Fig. 6.23.

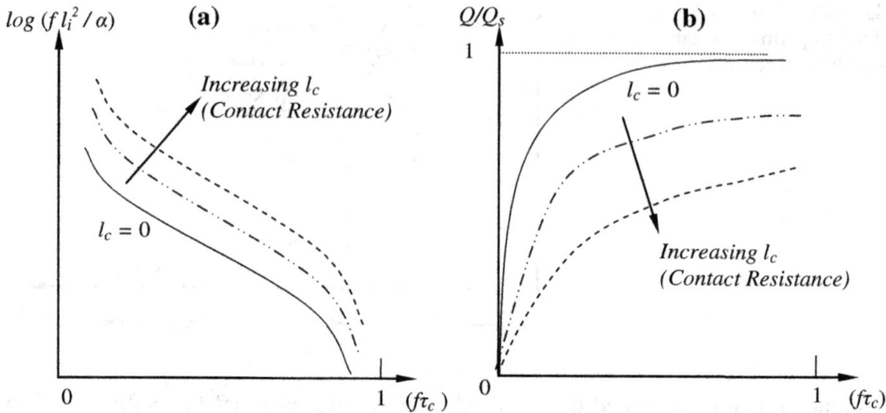

Fig. 6.21 Effect of contact resistance on: **a** the variation of disrupted heat flow with contact time; and **b** the variation of heat flow rate with contact time. Q is the actual heat flow and Qs is the steady state heat flow

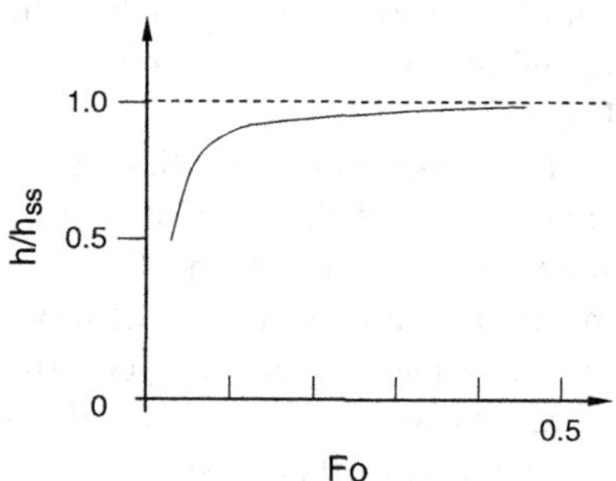

Fig. 6.22 Variation of contact conductance with time (after Moses and Johnson 1988)

The data represent results of seven tests: five on specimens of brass, and one each of copper and aluminium. It is clear that the larger the contact time and/or the conductance, the shorter is the time needed to achieve the quasi-steady state.

Moses and Johnson, in fact, found that a much smaller number of cycles was needed to *approach* (for example within 15 % of) the steady state conditions. These are also shown plotted in the same graph.

In a subsequent paper, Moses and Johnson (1989) noted that, for fixed values of contact conductance, changes in contact time τ_c alter the temperature distribution at the end of the separation part of the cycle, but they do not significantly modify the temperature distribution during contact. These results were confirmed by Moses and Dodd (1990). They also confirmed that the thermal contact conductance was not constant throughout the contact period. For a given set of experimental parameters, the ultimate value of contact conductance depended only on the contact time τ_c.

Although exact temperature distributions in the regions of temperature fluctuation can be determined by sophisticated analytical or numerical calculations, if

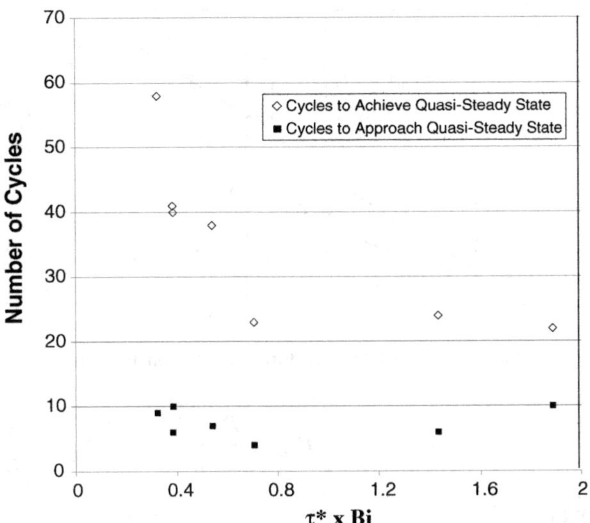

Fig. 6.23 Number of cycles needed to achieve or approach quasi-steady state (plotted from the data of Moses and Johnson 1988)

the main concern is to evaluate the cycle averaged heat flux, then the following procedure proposed by Wiŝniewski 1994 (see also Furmañski and Wiŝniewski 2002) may be used. Wiŝniewski used "properly defined thermal resistances" which depend upon the surface temperatures T_{1E} and T_{2E} extrapolated from the linear temperature distributions within the solids (see Fig. 6.24) to investigate the periodic contact problem.

During at least part of the contact period, the TCR is of an unsteady nature and its value can vary because of the following reasons:

- the difference in surface temperatures immediately before contact is higher than the temperature difference after contact
- the interface pressure changes during loading and unloading due to vibrations, collisions and rebounds of the solids
- the temperature in the interface region varies during contact.

Therefore it is very difficult to measure TCR at periodic contact, particularly for very short periods. The only reasonable way to make periodic contact heat transfer calculations seems to apply (or make use of) the TCR measured during steady contact in order to estimate the real value of the total resistance.

Referring to Fig. 6.24, the steady state contact resistance is:

$$R_{cs} = \frac{T_2^0 - T_1^0}{q_s} \tag{6.36}$$

where q_s is the heat flux at steady (permanent) contact.

Since the TCR during periodic contact, used in the calculation of the total resistance, must be related to the cycle time, it should be defined by the following relation:

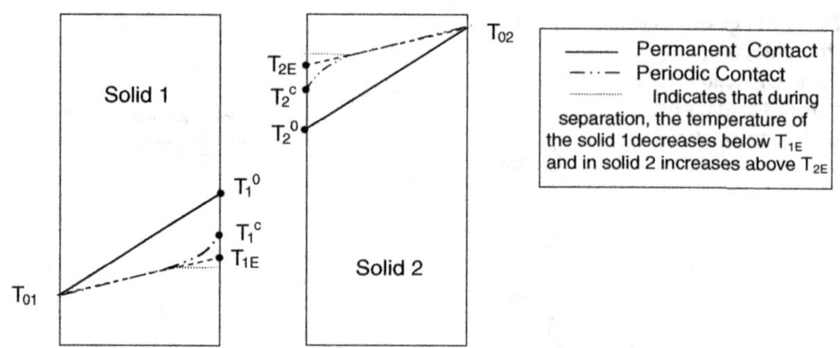

Fig. 6.24 Example of temperature distribution in solids

$$R_{cp}^{*}=\frac{T_2^c-T_1^c}{q_p}=\frac{\tau}{\tau_c}R_{cp}\cong\frac{\tau}{\tau_c}R_{cs} \tag{6.37}$$

where

R_{cp} = TCR during the periodic contact
R_{cp}^{*} = TCR during the periodic contact referred to cycle time
$T_{1,2}^c$ = surface temperatures during periodic contact
q_p = mean value of the heat flux over cycle time
τ = cycle time (sum of separation and contact times)
τ_c = contact time

As explained earlier, the interruption of heat flow causes additional thermal resistance to appear in each solid, which will be called the *periodic contact resistance* (PCR). In the case of perfect thermal contact, and for adiabatic conditions on the contacting surfaces during separation, this would be the only thermal resistance present between periodically contacting solids. The PCR depends on the contact time, the separation time, thermal diffusivity and the thermal conductivity of each solid. The PCR's, designated, R_{p1} and R_{p2}. for solids 1 and 2 respectively, are defined by using the extrapolated temperatures T_{1E} and T_{2E}:

$$R_{p1}=\frac{T_1^c-T_{1E}}{q_p}=\frac{1}{h_{p1}};\quad R_{p2}=\frac{T_{2E}-T_2^c}{q_p}=\frac{1}{h_{p2}} \tag{6.38}$$

where h_{p1} and h_{p2} are the periodic contact conductances.

The total thermal resistance is then given by

$$R_T=R_{p1}+R_{cp}^{*}+R_{p2} \tag{6.39}$$

This equation assumes that adiabatic conditions exist on the contacting surfaces during separation.

As indicated earlier, the determination of PCR's requires analytical or numerical calculations. The results of such calculations may be generalised using dimensionless parameters P, defined for each solid as:

$$P_1 = \frac{h_{p1}}{k_1}(\alpha_1 \tau)^{0.5}; P_2 = \frac{h_{p2}}{k_2}(\alpha_2 \tau)^{0.5} \tag{6.40}$$

where k and α are the thermal conductivity and diffusivity, respectively.

These non-dimensional parameters are independent of the thermal properties of the solids; they only depend on the quantities characterising the periodic contact, namely, the cycle period and the proportion of the cycle time spent in contact. Therefore one common graph can be made for all solids (see Fig. 6.25). Further, in a given pair of solids, each solid has the same cycle time and the same contact time and hence

$$P_1 = P_2 = P$$

Consequently, when the values of the cycle time and the contact time are known, the value of P may be read from the chart. Then using the definition of P, the periodic contact conductance values h_{p1} and h_{p2} may be determined.

Shojaefard et al. (2008) conducted an analytical and experimental study to estimate the average thermal resistance of the surfaces that are making intermittent contact, given the frequency and the proportion of the cycle time spent in contact.

During experimentation, two co-axial rods were used to transfer heat at their contact surfaces. Using the measured temperatures at different locations on the rods and the analytical solution, the temperature distribution of the rods and the heat transfer coefficient of the contact surface were calculated. Using the above calculated temperatures at both sides of the contact surface and applying the system identification method, the temperature transfer function was estimated. By

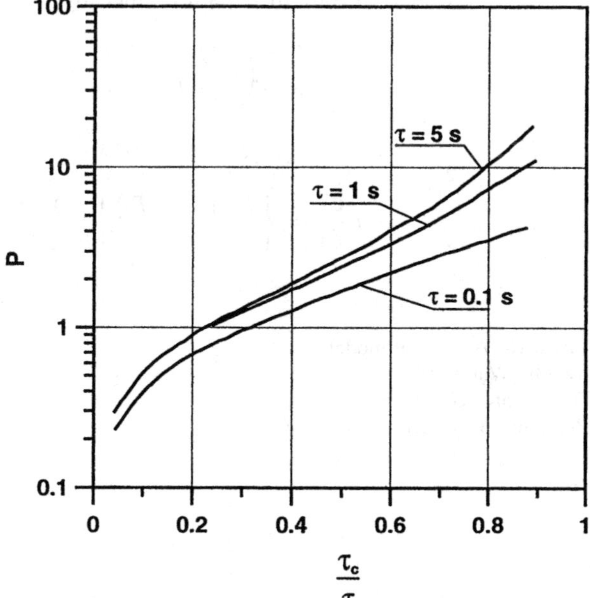

Fig. 6.25 Dimensional parameter P versus the proportion of cycle time spent in contact (τ_c/τ), (Figure reproduced by permission from Professor Wiŝniewski)

using the calculated transfer function, a computational model can be created. The computation model can then be used, for example, to estimate the exhaust valve temperature (a notoriously difficult thing to measure) when the exhaust valve seat temperature is used as an input, is extremely difficult to measure in practice. *Note: The transient heat transfer experiment was conducted with the test specimens (cylinders) in constant, not intermittent, contact.*

Wang and Degiovanni (2002) developed a quadrupoles method has been developed to solve heat transfer through a macro-contact with thermal constriction, which is periodic in time and two-dimensional in space. The constriction was taken into account by means of a constriction term present in the quadrupole matrix.

The quadrupole method is based on 2 × 2 matrices that relate some *transform* (in the present work, Laplace Transform) of both temperature and flux on one surface of a medium under consideration to the same quantities on another surface.

The analytical model consisted of a cylindrical rod of length l with a uniform cross-section of radius R which is insulated laterally so that no heat transfer takes place from the sides. The thermal conductivity of the rod is k, and its diffusivity α. One end ($x = 0$) is held at a uniform temperature T_0 while the other end ($x = l$) is brought into periodic contact (contact–noncontact) with a plane kept at constant temperature T_c. This contact involves a disk (the asperity) of radius $r_0 < R$ (Fig. 6.26). It is assumed that:

- the contact conductance is time dependent and uniform over the whole asperity;
- there is no heat transfer through the gap surrounding the asperity;
- the thickness of the asperity is negligible.

The heat transfer problem is then defined by:

$$\frac{1}{r}\frac{\partial}{\partial r}\left(r\frac{\partial T}{\partial r}\right)+\frac{\partial^2 T}{\partial x^2}=\frac{1}{\alpha}\left(\frac{\partial T}{\partial t}\right)$$

$$T = T_0 \text{ for } x = 0$$

$$-k\frac{\partial T}{\partial x}=\begin{cases} h(t)(T-T_0) & \text{for } 0<r<r_0, \\ 0 & \text{for } r_0<r<r, \end{cases} x = l$$

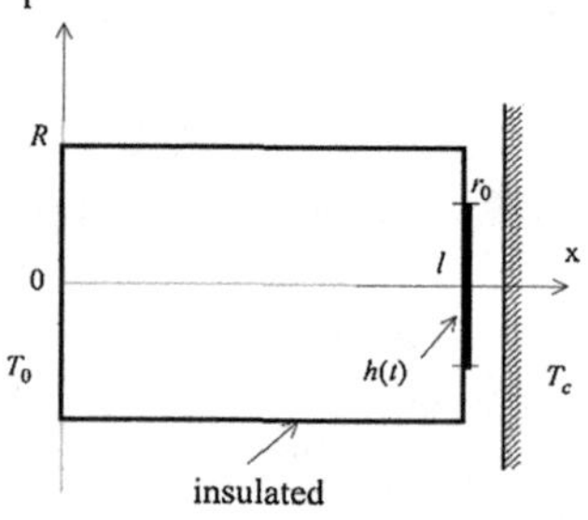

Fig. 6.26 Analytical model used by Wang and Degiovanni (2002); Reproduced by permission from Pergamon

$$\frac{\partial T}{\partial r} = 0 \text{ for } r = R$$

where the contact resistance per unit area 1/h(t) varies periodically with time in a period $\tau = \tau_c + \tau_0$. It is equal to $1/h_c$ during the contact phase τ_c and to $1/h_0$ during the noncontact phase τ_0.

The study showed that the one-dimensional periodic model remains valid for long contact periods, however it was necessary to introduce the concept of "building-up" of the constriction to explain the thermal contact behaviour for short and moderate periods.

Figure 6.27 shows the dimensionless constriction resistance, that is the ratio of the constriction resistance and its permanent value $r_{cts} = 8/3\pi^2 k r_0$, as a function of the dimensionless time $t^* = t/(r_0^2/\alpha)$. The constriction resistance approaches its permanent value at a particularly slow pace: 0.90 at $t^* = 10$; 0.97 at $t^* = 100$; 0.99 at $t^* = 1000$. In other words, the ''building-up'' time for 90 % of the constriction resistance is: $\tau_{ct} \approx 10 r_0^2/\alpha$. The characteristic time of the rod was defined as $\tau_b = l^2/\alpha$.

The authors identified three states for thermal resistance according to the relative magnitudes of the contact time, τ_c, the characteristic time of the rod τ_b and the characteristic time of the constriction, τ_{ct}.

- Large contact time, $\tau_c > \tau_b > \tau_{ct}$: The constriction and the thermal field are quasi-steady. It corresponds to the situations of contact during phase τ_1 and noncontact during τ_2; the solution of the problem is simply the addition of two steady states.
- Moderate contact time, $\tau_b > \tau_c > \tau_{ct}$: The contact period is sufficiently short so that the thermal field in the rod has no time to evolve except in the constriction zone; the thermal field in the bar behaves then as in steady state, only the constriction resistance varies periodically with time.
- Short contact time, $\tau_b > \tau_{ct} > \tau_c$: The contact period is so small that, neither in the rod nor in the constriction zone, does the thermal field have time to evolve; the

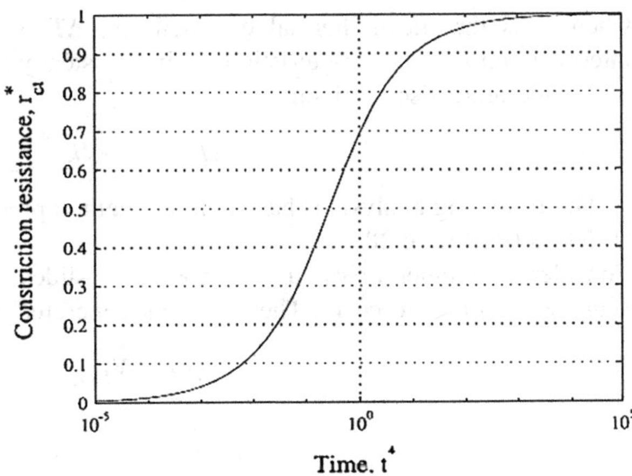

Fig. 6.27 Constriction resistance vs time Wang and Degiovanni (2002); Reproduced by permission from Pergamon

system resistance is then the same as in steady state with an average resistance at the interface equal to the reciprocal of the average conductance.

Periodic Contacts—Summary. The theoretical works reviewed above represent different approaches to the analysis of the problem. The works of Howard and Sutton, discussed here, put a limitation on the length of the rod and also require that the two bars be of the same material-whereas the theory of Wisniewiski imposes no such restrictions. Both of these theories need that the contact resistance be separately estimated. On the other hand, the model of Wang and Degiovanni use the resistance associated with a single constriction rather than the contact resistance. The experimental works indicate that (a) the quasi steady state is quickly *approached* in a periodic contact (typically less than 10 cycles), although it may take longer to achieve it, and (b) the contact conductance during the contact time cannot be assumed to be a constant, especially if the duration of contact is small.

6.4 Thermal Contact Resistance and Sliding Friction

The generation of frictional forces between the sliding bodies is fundamentally an energy dissipating mechanism, and this gives rise to heating effects which originate at the sliding interface.

If f is the coefficient of friction, P is the normal pressure on the contacting surfaces and V is the relative velocity, then the heat generated per unit area due to friction is:

$$q = fVP \tag{6.41}$$

The heat conducted away from the interface is

$$q_c = k\Delta T/L \tag{6.42}$$

where k is the mean thermal conductivity, ΔT is the temperature rise of the interface and L is a characteristic length. For steady state conditions, $q = q_c$, and the temperature rise is given by

$$\Delta T = fVPL/k \tag{6.43}$$

The following analysis is based on the method presented by Bowden and Tabor (1958, 2001) Fig. 6.28.

Consider a cylinder of radius r, whose face slides over a surface with velocity V under a normal force W. Then the heat generated due to friction is

$$Q = fWV \tag{6.44}$$

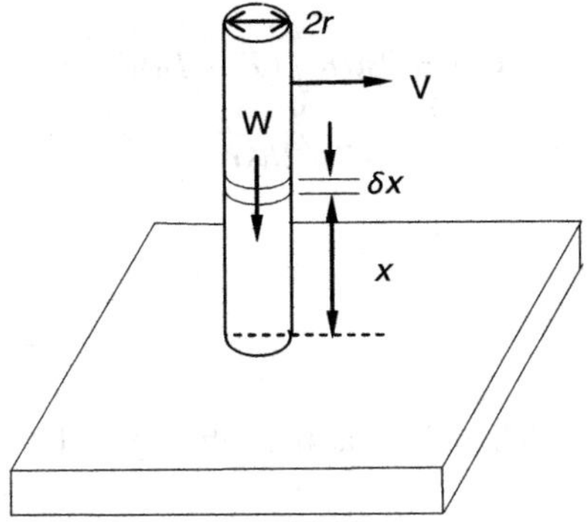

Fig. 6.28 Cylinder end sliding over a flat surface

Assuming that the interface is at a constant temperature, and that the heat flow along the cylinder is one-dimensional, the heat gained by the element of thickness δx is

$$\frac{d}{dx}\left(kA\frac{dT}{dx}\right)\delta x = kA\frac{d^2T}{dx^2}\delta x$$

where $A = \pi r^2$ is the area of cross-section of the cylinder and it is assumed that k is independent of temperature.

The heat loss from the element to the surroundings (at temperature T_0) is

$$h(2\pi r)\delta x(T - T_0)$$

where h is the (convection + radiation) heat transfer coefficient.

For steady state conditions, these two should be equal

$$k(\pi r^2)\frac{d^2T}{dx^2}\delta x = h(2\pi r)\delta x(T - T_0) \tag{6.45}$$

Or

$$\frac{d^2T}{dx^2} - \frac{2h}{kr}(T - T_0) = 0$$

This has the solution

$$T - T_0 = Be^{-\left(\sqrt{2h/kr}\right)x} \tag{6.46}$$

B is a constant to be determined.

Some of the frictional heat will go to the lower body and the remaining to the cylinder.

If C is the fraction of Q that goes to the cylinder, then this is the amount that will be lost to the surroundings. Therefore, assuming the cylinder to be very long,

$$CQ = 2\pi rh \int_0^\infty (T - T_0)dx = 2\pi rh \int_0^\infty Be^{-\left(\sqrt{2h/kr}\right)x}dx = 2\pi rhB\left(\frac{1}{\sqrt{2h/kr}}\right)$$
$$= \pi rB\sqrt{2hkr}$$

Hence

$$T - T_0 = \frac{CQ}{\pi r\sqrt{2hkr}} e^{-\left(\sqrt{2h/kr}\right)x} \tag{6.47}$$

Therefore, at the rubbing surface, where $x = 0$, the temperature rise is given by

$$T - T_0 = \frac{CfWV}{\pi r}\left(\frac{1}{\sqrt{2hkr}}\right) \tag{6.48}$$

Numerical Example:

Consider a constantan cylinder of 1 mm diameter, sliding over a mild steel surface with a velocity of 2 m/s, under a force of 1 N. It is known that the coefficient of kinetic friction $f = 0.3$, the conductivity of constantan is 23 W/(mK), and the heat transfer coefficient $h = 42$ W/(m^2 K). Assuming that half of the heat generated goes into the cylinder, the temperature rise in the constantan surface will be:

$$T - T_0 = \frac{0.5 \times 0.3 \times 1 \times 2}{\pi \times 0.5 \times 10^{-3}}\left(\frac{1}{\sqrt{2 \times 42 \times 23 \times 0.5 \times 10^{-3}}}\right) = 194\,^{\circ}\mathrm{C}$$

Note: The above calculation assumes that contact occurs over the whole end face of the cylinder. But since the actual area of contact is only a very small fraction of the apparent area, it can be expected that the temperature rise in the small region of contact will be much higher.

When sliding with friction occurs between two solid surfaces, the heat is dissipated into the bulk of the solids. Therefore the heat flow does not depend on the rate at which the heat is lost to the environment—rather it depends on the thermal conductivities of the solids in contact. The following is a simple explanation of this process.

Consider two surfaces, 1 and 2, with thermal conductivities k_1 and k_2 touching over a small circular region of radius a. From the frictional heat generated, Q, let fraction Q_1 go to body 1 and Q_2 to body 2, such that

$$Q = Q_1 + Q_2$$

At steady state, the junction will be at a temperature T, while the bulk of the solids remain at temperature T0. Since the constriction resistance of the circular area is 1/(4*ak*), we have

$$Q = Q_1 + Q_2 = 4ak_1(T - T_0) + 4ak_2(T - T_0)$$

Therefore

$$(T - T_0) = \frac{Q}{4a}\left(\frac{1}{k_1 + k_2}\right) \tag{6.49}$$

If f is the coefficient of friction, P the normal force and V is the sliding velocity, then

$$(T - T_0) = \frac{fWV}{4a}\left(\frac{1}{k_1 + k_2}\right) \tag{6.50}$$

A more rigorous analysis by Jaeger (1942), for a square junction of side $2L$, gave the following result:

$$(T - T_0) = \frac{fWV}{4.24L}\left(\frac{1}{k_1 + k_2}\right) \tag{6.51}$$

which is very similar to the approximate equation, except for the constant.

The above steady state analysis applies only to slow sliding speeds. At higher speeds, the interface may not reach a steady state value because of the continuously and rapidly oncoming cool surface. As a result, the temperature rise will be less than the steady state value. For a square junction of side $2L$, Jaeger obtained the following result:

$$(T - T_0) = \frac{\alpha_1^{1/2} fWV}{3.76L}\left(\frac{1}{1.125k_2\alpha_1^{1/2} + k_1\sqrt{LV}}\right) \tag{6.52}$$

where body 1 is the smooth surface, and body 2 the surface which carries the slider and α_1 the diffusivity of body 1. Equation (6.52) indicates that the temperature rise $(T - T_0)$ occurs less rapidly than the first power of the velocity.

6.4.1 Thermoelastic Instability

As we have just seen, the frictional heat generated is proportional to the contact pressure. But the heat carried away from the junction depends on the contact resistance which, in turn, depends on the contact pressure. This leads to a phenomenon called frictionally excited thermoelastic instability (TEI), see Barber 1969 and Yeo and Barber 1995. As an example of the many analyses dealing with TEI, we will present here some features of the theory as put forth by Ciavarella and Barber 2005.

Consider the rectangular block of height 2 h and length L in out-of-plane sliding contact with a rigid wall at $x = 0$ and subjected to a normal pressure P_0 at $x = L$ (Fig. 6.29). The rigid wall is maintained at a temperature T_1, the other end of the block (at $x = L$) is maintained at temperature T_2 and the sides of the bar are thermally insulated from the surroundings. The heat generated q and the temperature, T, of the block at the interface are both variable with respect to time and

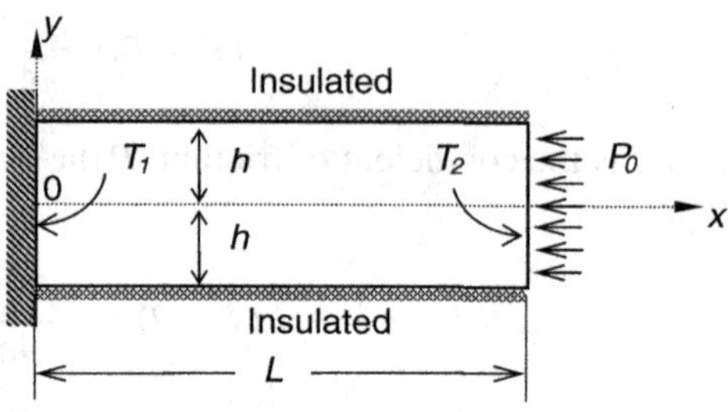

Fig. 6.29 The sliding contact model of Ciavarella and Barber 2005

the y-direction and the contact resistance, R, between the wall and the frictional heat source is dependent on the pressure.

The heat flow into the wall is:

$$q_w = (T - T_1)/R \tag{6.53}$$

and the heat flowing into the block is:

$$q_B = -k\frac{\partial T}{\partial x} = q - q_w = fVP - \frac{(T - T_1)}{R} \tag{6.54}$$

Steady State Conditions: According to Dundurs 1974, the steady state conduction of heat into the block at x = 0, will cause a locally convex curvature. Hence the steady state results in a non-uniform pressure with the maximum pressure at x = y = 0. However, sliding will cause local wear which is proportional to the contact pressure and this will ultimately tend to equalize the contact pressure, In the presence of wear, the only permissible steady state is the one in which a time-dependent wear rate leads to a kinematically admissible rigid-body motion. For the symmetrically loaded rectangular block, this requires that the wear rate and hence the steady-state contact pressure be uniform, that is, $p = p_O$. Also for steady state conduction, the temperature will be a linear function of x. Hence

$$\frac{\partial T}{\partial x} = \frac{(T_2 - T_0)}{L} \tag{6.55}$$

where T_0 is the steady state temperature of the bar at x = 0. From Eqs. (6.54) and (6.55) we get

$$T_0 = \frac{LR_0 fVP_0 + LT_1 + kR_0 T_2}{kR_0 + L} \tag{6.56}$$

and

$$q_0 = \frac{k(R_0 fVP_0 + T_1 - T_2)}{kR_0 + L} \tag{6.57}$$

in which $R_0 = R(P_0)$ and q_0 is the steady state value of q_B.

The stability of the steady state was analysed by performing a linear perturbation of Eq. (6.54) for q_B. For details, the reader is referred to the papers by Ciavarella and Barber 2005 and Yeo and Barber (1996). We will discuss the results here. Two cases were considered

1. *No frictional heating*

   Here the criterion for instability was found to be

$$E\alpha\gamma > 2(1-\nu)\left(1+\frac{3.88kR_0}{h}\right) \tag{6.58}$$

where E, α, and ν are the Young's modulus, the coefficient of thermal expansion and the Poisson's ratio respectively for the material of the rectangula block. γ is given by

$$\gamma = \frac{k(T_2-T_1)R'}{kR_0+L} \tag{6.59}$$

and R' is the derivative of the contact resistance R with respect to the pressure p.

2. *Effect of frictional heating*

   In this casc V ≠ 0 and the steady state with sliding becomes unstable if

$$E\alpha(\beta fV+\gamma) > 2(1-\nu)\left(1+\frac{3.88kR_0}{h}\right) \tag{6.60}$$

where

$$\beta = R_0 + \frac{LR'P_0}{kR_0+L}$$

Note that since the contact resistance decreases as the contact pressure increases, R' is usually negative. In the absence of sliding (case 1), the system is unstable only for sufficiently large negative values of (T_2-T_1), that is, when the heat flows from the wall into the block. With sliding present, instability can occur for lower negative values if and only if

$$\beta > 0$$

This condition is satisfied for most simple idealizations of the pressure law such as:

$$R_0 = B + \frac{A}{p_0}; A > 0, B > 0$$

When there is no externally imposed temperature difference, $(T_2-T_1)=0$, and, therefore, $\gamma = 0$. The stability criterion then becomes

$$E\alpha(\beta fV) > 2(1-\nu)\left(1+\frac{3.88kR_0}{h}\right) \tag{6.61}$$

Two significant conclusions can be noted from the above analysis:

1. The effect of sliding is to generally reduce the temperature difference required to cause instability.
2. The critical sliding speed depends on the contact pressure, in contrast to systems in which there is no TCR. This conclusion is somewhat expected since we started out with the premise that the instability is caused by the fact that the heat generated depends on the contact pressure and the heat carried away (being dependent on the contact resistance) also depends on the contact pressure.

References

Bolted or Riveted Joints

Aron W, Colombo G (1963) Controlling factors of thermal conductance across bolted joints in a vacuum environment. ASME Paper 63-HT-196, American Society of Mechanical Engineers, New York

Bradley TL, Lardner TJ, Mikic BB (1971) Bolted joint interface pressure for thermal contact resistance. Trans ASME, J Appl Mech, pp 542–545

Chandrashekhara K, Muthanna SK (1977a) Stresses in a thick plate with a circular hole under axisymmetric conditions. Int J Eng Sci 15:135–146

Chandrashekhara K, Muthanna SK (1977b) Axisymmetric stress distribution in a transversely isotropic thick plate with a circular hole. Acta Mech 28:65–76

Chandrashekhara K, Muthanna SK (1978) Pressure distribution in bolted connections. In: Advances in reliability and stress analysis, ASME, winter meeting. American Society of Mechanical Engineers, New York

Chandrashekhara K, Muthanna SK (1979) Analysis of a thick plate with a circular hole resting on a smooth rigid bed and subjected to axisymmetric normal load. Acta Mech 33:33–44

Curti G, Raffa F, Strona P (1985) Analysis of contact area and pressure distribution in bolted plates by boundary element method. Wire 35:14–18

Elliott DH (1965) Thermal conduction across aluminum bolted joints. ASME Paper 65-HT-53, American Society of Mechanical Engineers, New York

Fletcher LS, Peterson GP, Madhusudana CV, Groll E (1990) Constriction resistance through bolted and riveted joints. Trans ASME, J Heat Trans 112:857–863

Fernlund I (1961) A method to calculate pressure between bolted or riveted plates. Chalmers University of Technology, Gothenburg, pp 13–51

Gould HH, Mikic BB (1972) Areas of contact and pressure distribution in bolted joints. Trans ASME, J Eng Ind 94:864–870

Greenwood JA (1964) The elastic stresses produced in the mid-plane of a slab by pressure applied symmetrically at its surface. Proc Cambridge Phil Soc 60:159–169

Ito Y, Toyoda J, Negata S (1979) Interface pressure distribution in a bolt-flange assembly. Trans ASME, J Mech Des 101:330–337

Kumano H, Sawa T, Hirose T (1994) Mechanical behaviour of bolted joints under steady heat conduction. Trans ASME, J Press Vess Technol 116:42–48

Lardner TJ (1965) Stresses in a thick plate with axially symmetric loading. Trans ASME, J Appl Mech 87:458–459

Lee S, Song S, Moran KP, Yovanovich MM (1993) Analytical Modelling of Thermal Resistance in Bolted Joints ASME, HTD. Enhanc Cool Tech Electron Appl 263:115–122

Madhusudana CV (1994) Conduction of heat through bolted joints. In: Thermal conductivity 22, Technomic Publishing, Lancaster, PA, pp. 701–711

Madhusudana CV, Peterson GP, Fletcher LS (1990a) Effect of non-uniform pressures on the thermal conductance in bolted and riveted joints. Trans ASME, J Energy Res Technol 112:174–182

Minakuchi Y (1984) Contact stresses between two thick plates with a circular hole and a sandwiched solid metal gasket. Bulletin JSME 27:17–23

Minakuchi Y, Koizima T, Shibuya T, Yoshemine K (1983) Contact stresses in two elastic bodies (elastic bodies are finite hollow cylinders). Bulletin JSME 26:716–723

Mittlebach M, Vogd C, Fletcher LS, Peterson GP (1993) The interfacial pressure distribution and thermal conductance of bolted joints. ASME, HTD, Heat Transf Hazard Waste Process 212:9–17

Motosh N (1976) Determination of joint stiffness in bolted connections. Trans ASME, J Eng Ind 91:858–861

Oehler SA, McMordie RK, Allerton AB (1979) Thermal contact conductance across a bolted joint in a vacuum. AIAA Paper 79–1068. American Institute of Aeronautics and Astronautics, New York

Roca RT, Mikic BB (1972) Thermal conductance in a bolted joint. Prog Astro Aero 31:193–207

Rötscher F (1927) Die Maschinenelemente. Springer, Berlin, pp 230–236

Sneddon IN (1946) The elastic stresses produced in a thick plate by the application of pressure to its free surfaces. Proc Cambridge Phil Soc 42:260–271

Veilleux E, Mark M (1969) Thermal resistance of bolted or screwed sheet metal joints in a vacuum. J Spacecraft 6(3):339–342

Whitehurst CA, Durbin WT (1970) A study of thermal conductance of bolted joints. Report No. NASA-CR-102639, Louisiana State University, Baton Rouge

Yeh CL, Wen CY, Chen YF, Yeh SH, Wu CH (2001) An experimental investigation of thermal contact conductance across bolted joints. Exp Thermal Fluid Sci 25:349–357

Yip FC (1972) Theory of thermal contact resistance in vacuum with an application to bolted joints. Prog Astro Aero 31:177–192

Cylindrical Joints

Ayers GH, Fletcher LS, Madhusudana CV (1997) Thermal contact conductance of composite cylinders. J Thermophysics and Heat Transfer 11:72–81

Egorov ED, Nekrasov MI, Pikus VYu, Daschyan AA (1989) Investigation of contact heat transfer resistance in two-layer finned tubes. Soviet Energy Technol 6(6):33–37

Hsu TR, Tam WK (1979) On thermal contact resistance of compound cylinders. AIAA 14th Thermophysics Conf, Paper 79–1069. American Institute of Aeronautics and Astronautics, New York

Madhusudana CV (1999) Thermal conductance of cylindrical joints. Int J Heat and Mass Transfer 42:1273–1287

Madhusudana CV, Fletcher L, Peterson GP (1990b) Thermal conductance of cylindrical joints—a critical review. J Thermophys Heat Transfer 4:204–211

Popov VM, Krasnoborod'ko AI (1975) Thermal contact resistance in a gaseous medium. Inzhenerno-Fizicheski Zhurnal 28(5):875–883

Sheffield JW, Stafiord BD, Sauer Jr, HJ (1984) Finned tube contact conductance: investigation of contacting surfaces. ASHRAE Trans, 90:442–452

Sheffield JW, Abu-Ebid M, Sauer Jr, HJ (1985) Finned tube contact conductance: empirical correlation of thermal conductance. ASHRAE Trans, 91 (2a): 100–117

Sheffield JW, Sauer HJ Jr, Wood RA (1987) An experimental method for measuring the thermal contact resistance of plate finnned tube heat exchangers. ASHRAE Trans 93(2):776–785
Sheffield JW, Wood RA, Sauer Jr, HJ (1989) Experimental investigation of thermal conductance of finned tube contacts. Exp Therm Fluid Sci 2:107–121
Timoshenko (1956) Advanced Strength of Materials, Third Edition, p 211
Timoshenko SP, Goodier JN (1970) Theory of Elasticity, 3d edn. McGraw-Hill, New York, pp 443–452
Tsukizoe T, Hisakado T (1965) On the mechanism of contact between metal surfaces—the penetrating depth and the average clearance. Trans ASME, J Basic Eng 87:666–674
Williams A, Madhusudana CV (1970) Heat flow across cylindrical metallic joints. 4th international heat transfer conference, Paper Cu 3.6. Elsevier, Amsterdam, The Netherlands

Periodic Contacts

Eckert ERG, Drake RM (1959) Heat and mass transfer. McGraw Hill Book Company Incorporated, New York
Furmański P, Wiśniewski TS (2002) Thermal contact resistance and other thermal phenomena at soli-solid interface. Institute of Heat Engineering, Warsaw Institute of Technology, Warsaw, pp 160–166
Howard JR, Sutton AE (1970) An analogue study of heat transfer through periodically contacting surfaces. Int J Heat and Mass Transfer 13:173–183
Howard JR, Sutton AE (1973) The effect of thermal contact resistance on heat transfer between periodically contacting surfaces. Trans ASME, J Heat Transfer 95:411–412
Moses WM, Johnson RR (1988) Experimental study of the transient heat transfer across periodically contacting surfaces. J Thermophysics 2:37–42
Moses WM, Johnson RR (1989) Experimental results for the quasisteady heat transfer through periodically contacting surfaces. J Thermophys 3:474–476
Moses WM, Dodd NC (1990) Heat transfer across aluminium/stainless-steel surfaces in periodic contact. J Thermophys 4:396–398
Shojaefard MH, Ghaffarpour M, Noorpoor AR (2008) Thermal contact analysis using identification method. Heat Transfer Eng 29:85–96
Wang H, Degiovanni A (2002) Heat transfer through periodic macro-contact with constriction. Int J Heat and Mass Transfer 45:2177–2190
Wiśniewski TS (1994) The method of calculation of heat transfer between periodically contacting solids. J Polish CIMAC 1:303–309

TCR and Sliding Friction

Barber JR (1969) Thermoelastic instabilities in the sliding of conforming solids. Proc R Soc Lond A312:381–394
Bowden FP, Tabor D (1958 Reprinted 2001) The Friction and Lubrication of Solids. Oxford University Press, London, pp 33–52
Jaeger JC (1942) Moving Sources of Heat and the Temperature of Sliding Contacts. J ProcRoy Soc, New South Wales 76:203–224
Ciavarella M, Barber JR (2005) Stability of thermoelastic contact for a rectangular elastic block sliding against a rigid wall. Eur J Mech A/Solids 24:371–376
Dundurs J (1974) Distortion of a body caused by free thermal expansion. Mech Res Commun 1:121–124

Yeo T, Barber JR (1995) Stability of a semi-infinite strip in thermoelastic contact with a rigid wall. Int J Solids and Structures 32:553–567

Yeo T, Barber JR (1996) Finite element analysis of the stability of static thermoelastic contact. J Therm Stresses 19:169–184

Chapter 7
Control of Thermal Contact Conductance Using Interstitial Materials and Coatings

As noted in Chap. 1, the actual solid-to-solid contact area, in most mechanical joints, is only a small fraction of the apparent area. The voids between the actual contact spots are usually occupied by some conducting substance such as air. Other interstitial materials may be deliberately introduced to control, that is, either to enhance or to lessen, the TCC: examples include foils, powders, wire screens and epoxies. To enhance the conductance the bare metal surfaces may also be coated with metals of higher thermal conductivity by electroplating or vacuum deposition. Greases and other lubricants also provide alternative means of enhancing the TCC.

In this chapter, interstitial materials will be grouped into the following categories for a discussion of their effect on TCC.

- Solid interstitial materials including phase change materials
- Metallic and other coatings
- Carbon black and carbon nanotubes
- Thermal greases, pastes and lubricant films
- Insulating interstitial materials

7.1 Solid Interstitial Materials

Filler materials, in general, make the joint less sensitive to mechanical loading and surface conditions. Therefore, in addition to providing a means to control the TCC of a joint, the fillers also make the joint behaviour more predictable. Comprehensive reviews of the general role of interstitial materials in the control of TCC have been published from time to time (see, for example, Snaith et al. 1984; Sauer 1992; Madhusudana et al. 1996; Madhusudana and Villanueva 1996; Prasher et al.

An erratum to this chapter is available at 10.1007/978-3-319-01276-6_11

C. V. Madhusudana, *Thermal Contact Conductance*,
Mechanical Engineering Series, DOI: 10.1007/978-3-319-01276-6_7,

2004). Some reviews have dealt with specific applications, for example, space craft thermal control (Fletcher 1973) and Gwinn and Webb 2003—Thermal challenges in Next Generation Electronic Systems.

Because the TCC is directly proportional to the thermal conductivity and inversely proportional to the hardness, a general rule to be followed for the enhancement of TCC (either with the use of foils or by coating the surfaces) is (see Snaith et al. 1984):

$$\frac{H_s k_c}{H_c k_s} > 1 \tag{7.1}$$

in which the subscript s refers to substrate and the subscript c refers to coating (or foil).

When interstitial materials are used for the control of thermal conductance, it is convenient to define an effectiveness e, to indicate their relative worth:

$$e = \frac{h_{cm}}{h_{bj}} \tag{7.2}$$

in which the subscripts cm and bj refer to with control material and bare joint respectively. This section reviews the role that the interface materials have played in controlling the TCC over the last 50 years. For those categories in which a significant amount of experimental data is available, tables are presented summarising their results. It must be noted that the data presented is, by no means, exhaustive; only representative values are tabulated. It should also be noted that the figures given in these tables are approximate due to the following reasons: In some cases, they are extracted from graphs; in some cases they have been converted from imperial units; and in some instances the pressure and conductance per unit area have been calculated from the contact force, the total conductance and the size of test specimens as published in the relevant document.

7.2 Metallic Foils

In situations where the mechanical load has to be limited because of design considerations and/or when the joint is in a vacuum, metal foils may be sandwiched between the bare metal surfaces. It would then be expected that the foils will fill the gaps between the surfaces thus increasing the actual contact area and enhancing the conductance.

One of the earliest experimental investigations into the effect of metal foils was conducted by Fried and Costello (1962). They used foils of lead and aluiminium between aluminium 2024-T3 surfaces.

On the other hand, Cunnington's (1964) experiments compared the conductance of bare aluminium 6061-T4 junction into which an indium foil had been inserted. Typical results (recast into SI units) of both of these works are

Table 7.1 Effect of surface texture on contact conductance with a filler material

Reference	Foil and foil thickness (μm)	Contact pressure (kPa)	Surface roughness (μm) rms	Flatness deviation (μm)	Contact conductance (W/m² K) bare junction	Contact conductance (W/m² K) with foil
Fried and Costello (1962)	Lead (200)	69	1.50–2.00	1125	190	730
		207			300	830
Fried and Costello (1962)	Aluminum (50)	69	12–16	425	265	390
		207			465	665
Cunnington (1964)	Indium (25)	276	0.30–0.45	0.875	2,270	13,060
			1.15–1.25	0.625	2,780	21,580
		552	0.30–0.45	0.875	3,520	18,740
			1.15–1.25	0.625	4,540	27,260

summarised below, in Table 7.1, in order to highlight the effect of surface texture of the contacting surfaces. In each case, the bare junction conductances were measured in vacuums of less than 10^{-4} Torr. It may be noted that both of these works also dealt with interstitial materials other than foils. These will be discussed at the appropriate sections.

An inspection of this table reveals that:

1. The TCC of a bare joint in vacuum is significantly increased by the insertion of a metallic foil.
2. Softer materials such as lead and indium were more effective in enhancing the conductance.
3. Cunnington's tests show that the foil was more effective in enhancing the conductance of the rougher surfaces indicating that the optimum thickness depends on the roughness of the surfaces. In his tests, it appears that the thickness of the foil was closer to the optimum required for the rougher of the two pairs of surfaces.

Another early experimental investigation into the effect of interfacial foils was that of Koh and John (1965). In their test, foils of copper, aluminium, lead, and indium were separately tested as TIM's between a pair of mild steel surfaces. Although copper and aluminium have high thermal conductivities, it was found that the insertion of these foils actually reduced the thermal contact conductance, whereas the lead and indium foils contributed toward an increase in the conductance. It was therefore confirmed that foil softness was more important than foil conductivity. In another series of tests, the same authors found that there was an optimum thickness of foil, which would result in maximum enhancement of joint conductance. Apparently, thick foils are not pliable enough to fill the voids in the joint, while too thin a foil may not provide sufficient conduction material to fill the gaps in the interface. For the surface roughness range of 4–5 μm rms encountered in the tests, the optimum thickness was found to be about 25 μm; at this thickness, the value of the conductance was about three times that for the bare junction. It was also found that, for the specimens tested, a foil thickness greater than 100 μm produced no improvement in the TCC.

Yovanovich (1972) conducted a detailed experimental study of the effect of lead, tin, aluminium and copper foils on the conductance of Armco iron joints. The contact pressure ranged from 2 to 10 MPa and the foil thicknesses from 20 to 500 μm. Unlike the tests of Fried and Costello, and those of Cunnington, the tests were conducted in atmospheric conditions. He found that an optimum thickness, corresponding to minimum joint resistance, existed in all cases. The ratio of optimum thickness to surface roughness (rms) was found to be about 2 for lead, between 0.48 and 0.58 for aluminium and 0.68 for copper. It was proposed that a foil material may be ranked by the ratio of its thermal conductivity to hardness; the larger the ratio, the greater will be the increase in contact conductance.

The investigations of O'Callaghan et al. (1983) and O'Callaghan and Probert (1988) led them to conclude that, in the absence of macroscopic constrictions, the optimum film thickness should be of approximately the same magnitude as the separation between the mean planes of solid surfaces. Thus, for, nominally flat surfaces, a relatively thin foil would be sufficient to enhance the conductance.

For surfaces with macroscopic errors of form, or when macroscopic thermal distortions are expected, the thickness should be larger in order to bridge any gaps that would be formed. The first of these conclusions indicates that the surfaces tested by Yovanovich were nominally flat. The second conclusion confirms the results of Fried and Costello (Table 6.1), who used a very thick lead foil to maximise the conductance of a pair of surfaces with large flatness deviation.

An example of the extensive experimental results of Peterson and Fletcher (1988) on the TCC in the presence of foils is shown plotted in Fig. 7.1a. In each case, the thickness of the foil corresponded to the optimum value suggested by Yovanovich (1972). Their tests also indicated that the enhancement of the conductance can be accurately ranked using the ratio of the foil thermal conductivity to its hardness; the higher the value, the greater is the enhancement. Indeed, Madhusudana (1994) showed that the four separate graphs showing the results for lead, tin, aluminium and copper may be reduced to virtually a single line if the conductance and the contact pressure values were normalised by dividing them by foil conductivity and foil hardness, respectively (see Fig. 7.1b). Peterson and Fletcher also noted that very thin foils (corresponding to the optimum thickness required for flat smooth surfaces) were difficult to handle. Consequently, if they were not able to be applied correctly, an actual decrease in conductance occurred due to the unintentional creation of folds and wrinkles.

Villanueva (1997) conducted tests on aluminium and gold foils inserted in joints formed between Nilo 36 (a low-expansion alloy similar to invar) and Stainless Steel (AISI 304). The tests were conducted in a vacuum of 1.5×10^{-3} Torr. His results for gold foils only are discussed here. The relevant properties of the parent metals and the gold foil are listed in Table 7.2. The effective roughness of the contacting surfaces is calculated to be 0.927 μm.

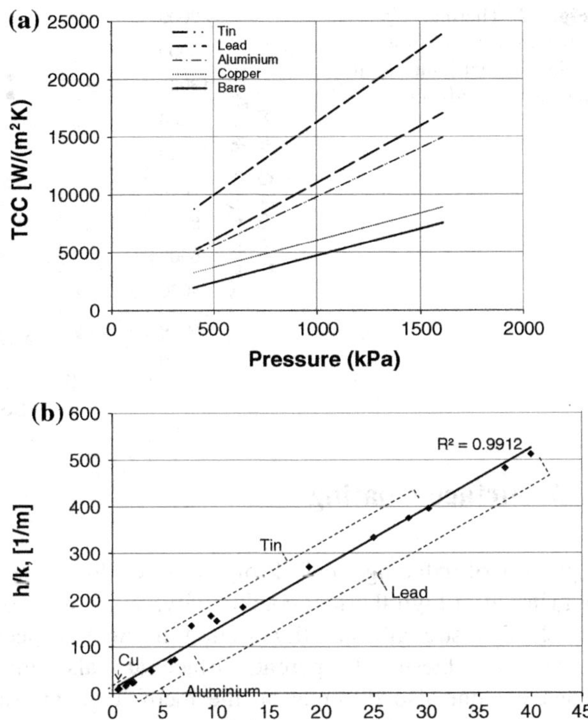

Fig. 7.1 Thermal contact conductance in the presence of foils. **a** Data of Peterson and Fletcher (1990); **b** "normalised data"

Table 7.2 Properties of specimens and gold foil used (Villanueva 1997)

Specimen material	Roughness (rms), μm	Slope (rms), radians	Thermal conductivity, W/(mK)	Microhardness, MPa
SS 304	0.5010	0.0717	14.8	3560
NILO 36	0.7800	0.1217	14.2	2760
Gold foil	0.15 (thickness)		315	294

The results are shown plotted in Fig. 7.2. It is seen that the conductance increased with the contact pressure. Also the conductance increased continuously with increasing number of gold leaves used. The equivalent thickness of five leaves of gold is 0.750 μm. This means that the optimum thickness was not reached in the tests. This was to be expected since the equivalent thickness of five leaves of gold is 0.750 μm which is less than the effective roughness of the surfaces in contact. It may also be noted that there was relatively less enhancement when only one gold leaf was used. This was probably due to the tearing of the very thin film resulting in several regions of bare metal-to-metal contact.

Table 7.3 provides a summary of representative experimental results for the enhancement of TCC in the presence of metallic foils.

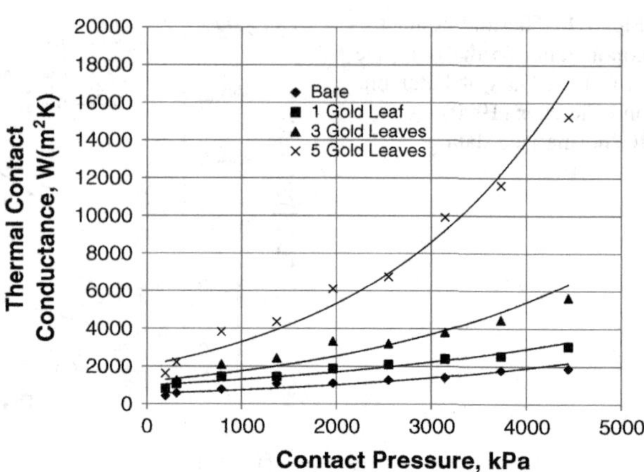

Fig. 7.2 Thermal enhancement of Nilo/stainless steel joint due to interstitial gold foil

7.3 Surface Coatings

One way of reducing the TCR of a joint would be to plate or coat the surfaces with a material of high thermal conductivity. Whenever coatings are contemplated, the mechanical strength, stability or durability with respect to operating conditions and time and adhesion to parent surface are also important considerations. The enhancement also depends on the method of deposition. For example, on aluminium alloy surfaces ($\sigma = 1\text{–}2\ \mu m$), electroplated silver coating of 12.7 μm thick, yielded a thermal enhancement of about 2.5 while a flame-sprayed silver coating of the same thickness on similar surfaces yielded an enhancement of only about 0.6–0.7 (Marotta et al. 1994). Plating would also result in a change in contact geometry for a given load since (a) the plating material might have a different flow pressure from that of the bare metal and, (b) the surface characteristics such as roughness and slope will change because of the plating (Madhusudana et al. 1996). For maximum benefit, both surfaces must be coated; when only one surface is coated, the whole constriction (or spreading) has to still take place in the other uncoated material.

7.3.1 Constriction Resistance in Plated Contact

The analytical model, proposed by Mikic and Carnasciali (1970) for a single plated contact is shown in Fig. 7.3.

Here b_I is a hypothetical intermediate radius. Physically, $(\pi b_I)^2$ represents the area at $z = t$, which is effectively used to transfer heat from the plated zone to the base. To determine b_I, it was noted that, for given boundary conditions of all possible resistances which may be obtained by choice of flow distributions, the actual resistance is the one that gives the minimum value. Hence for a fixed

Table 7.3 Thermal contact conductance in the presence of metallic foils

Details	Cunnington (1964)	Fried and Costello (1962)	Sauer (1992)	O'Callaghan and Probert (1988)	Peterson and Fletcher (1988)
Substrate material	*Aluminium 6061-T4*	*Aluminium 2024-T3*	*Aluminium 2024-T4*	*Duralumim/titanium alloy*	*Aluminium 6061-T6*
Conductivity	177	206	206	141/5.6	177
Hardness	1080	1600	1600	430 (Duralumin)	1080
Roughness	a. 0.30–0.45 b. 1.15–1.25	a. 0.15–0.20 b. 1.2–1.6	0.77	8.91/3.19	1.15/8.57
Flatness	a. 0.875 b. 0.625	a. 112.5 b. 42.5			
Foil material	*Indium*	*a. Lead* *b. Aluminium*	*Aluminium*	*Aluminium*	*a. Copper* *b. Aluminium* *c. Lead* *d. Tin*
Conductivity	80	a.33 b.206	206	206	a. 380 b. 206 c. 33 d. 60
Hardness	10	a. 40 b. 270	270	270	a. 800 b. 270 d. 53
Thickness	25	a. 200 b. 50	640	25	a. 40 b. 30 c. 75 d. 100

(continued)

Table 7.3 (continued)

Details	Cunnington (1964)	Fried and Costello (1962)	Sauer (1992)	O'Callaghan and Probert (1988)	Peterson and Fletcher (1988)
		a. 0.069–0.207			
Pressure	*0.276–0.552*	*b. 0.069–0.207*	*0.69–3.45*	*0.160–0.382*	*0.10–1.60*
$h_{bare,vac}$	a. 2270–3520 b. 3780–4540	a. 190–300 b. 265–465		178–382	800–7500
$h_{bare,air}$			2960–4750		
h_{foil}	a. 13060–18740 b. 21580–27260	a. 730–830 b. 390–665	8520–13060	764–3972	a. 1800–10000 b. 2200–16000 c. 2200–18000 d. 4000–25000
$\frac{h_{foil}}{h_{bare,vac}}$	a. 5.75–5.32 b. 7.76–6.00	a. 3.84–2.77 b. 1.47–1.43		4.29–10.39	a. 2.25–1.33 b. 2.75–2.13 c. 2.75–2.4 d. 5–3.33
$\frac{h_{foil}}{h_{bare,air}}$			2.88–2.75		

Note Conductivity in (W/m K), hardness and pressure in (MPa), surface roughness, flatness and foil thickness in (μm), and conductance in (W/m^2 K)

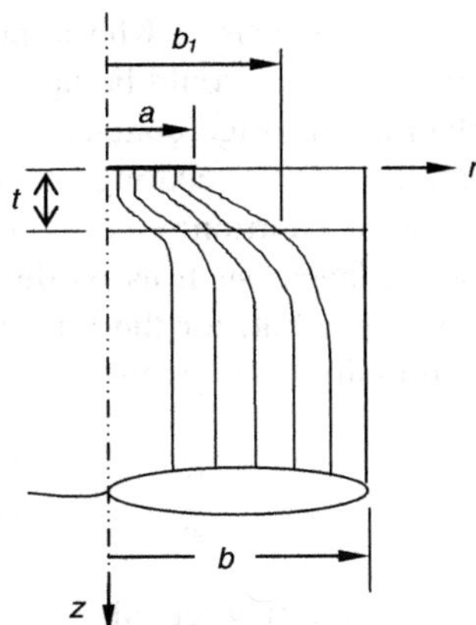

Fig. 7.3 A single plated contact

geometry and given thermal conductivities, the choice for b_1 for which the resistance is minimum, will yield the approximate (actually, an upper bound of) constriction resistance.

Let

R_t = resistance of the plated channel
R = resistance of the unplated channel

Then

$$R_t = \left(\frac{1}{4k_2 a}\right)\left[\left(\frac{16}{\pi C}\right)\varphi\left(\frac{t}{a},\frac{a}{b_1}\right)\right] + \left(\frac{1}{4k_2 b_1}F\left(\frac{b_1}{b}\right)\right) \tag{7.3a}$$

$$R = \left(\frac{1}{4k_2 a}\right)F\left(\frac{a}{b}\right) \tag{7.3b}$$

where

$C = k_1/k_2$
k_1= thermal conductivity of the plating material
k_2 = thermal conductivity of the base metal
F = the appropriate constriction alleviation factor
φ = the contact resistance factor determined according to the above procedure

The contact resistance factor will depend on the boundary condition, isothermal or constant flux, applied over the contact spot.

From the above equations, it is noted that

$$\frac{R_t}{R} = F\left(\frac{a}{b},\frac{t}{a},C\right) \tag{7.4}$$

A typical result is shown plotted in Fig. 7.4 for the conductivity ratio, $C = 5$. It is seen that considerable reduction in resistance can be achieved with sufficiently thick platings, $(t/a) > 2$.

The analysis of Kharitonov et al. (1974) also confirmed that the TCR of a flat rough surface could be noticeably altered when the thickness of coating is greater than the average contact spot radius, that is, $t > a$. Since a $\approx$ 30 μm, they concluded that the coatings of a few tens of microns thick would cause a significant change. Kharitonov et al. considered not only conducting coatings but also insulating layers such as oxide films. An open form solution was presented for the resistance but, for the limiting case when (b/a) $\rightarrow \infty$, the following approximate expression was given:

$$\frac{R_t}{R} = \frac{\left[1 + \frac{1}{C}\tanh\left(\frac{t}{ma}\right)\right]}{\left[1 + C\tanh\left(\frac{t}{ma}\right)\right]}; \quad \begin{cases} m = 1, \text{ for } k_1 < k_2 \\ m = 1.5, \text{ for } k_1 > k_2 \end{cases} \tag{7.5}$$

Another theoretical analysis of coated constrictions in half space was that due to Dryden (1983). In this analysis, the heat flux distribution over the contact spot was assumed to be given by

$$f(r) = \frac{-Q}{2k_1\pi(a^2 - r^2)^{0.5}}$$

As was seen in Chap. 2, this corresponds to an isothermal constriction. The heat conduction equation was solved using Hankel transform of the order zero. It was shown that the constriction resistance could be approximated by the following expressions.

Thin coatings, (t/a) < 2:

$$R_t = \left(\frac{1}{4k_2a}\right) + \left(\frac{1}{\pi k_1 a}\right)\left(\frac{t}{a}\right)\left(1 - C^2\right) \tag{7.6a}$$

Thick coatings, (t/a) > 2:

$$R_t = \left(\frac{1}{4k_2a}\right) - \left(\frac{1}{2\pi k_1 a}\right)\left(\frac{a}{t}\right)\left(\ln\frac{2}{1 + C}\right) \tag{7.6b}$$

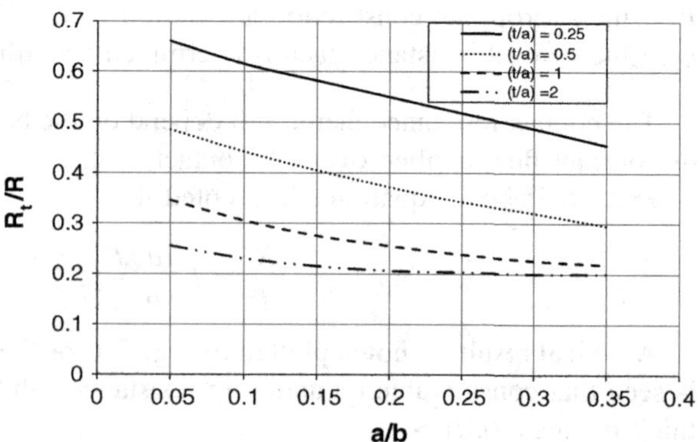

Fig. 7.4 Reduction of resistance due to plating. (based on the results Mikic and Carnasciali (1970) for C = 5; isothermal contact spot)

The analyses considered so far in this section refer to *plane circular* constrictions. A more realistic, plated *conical* constriction was analysed numerically by Mohs et al. (2000) and Olsen et al. (2001).

Plated constrictions in both vacuum and gaseous media were considered. Some typical results from Olsen et al. (2001) are shown in Fig. 7.5 a, b. In these diagrams:

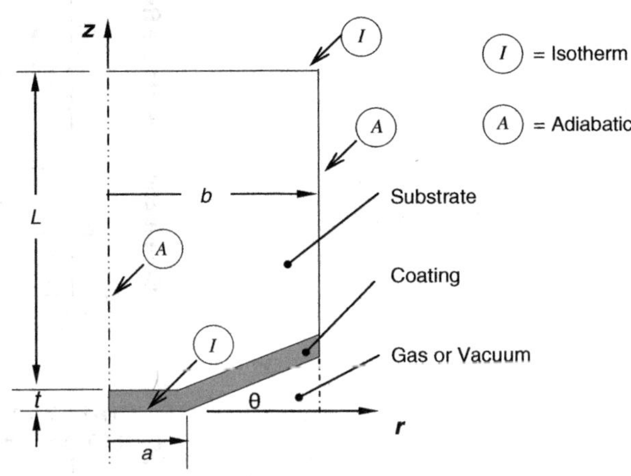

Fig. 7.5 Constriction resistance reduction versus thickness of coating. **a** Effect of varying the conductivity of the coating; **b** effect of varying the conductivity of the gaseous medium

k_s = thermal conductivity of substrate
k_c = thermal conductivity of coating material
k_g = thermal conductivity of gas (vacuum is indicated by $k_g = 0$)
g = temperature jump distance

Other quantities are defined in Fig. 7.6. The results showed that:

- For conductive coatings, the greatest reduction in the constriction resistance occurs at the smallest constriction ratio, although there is a reduction in resistance for all cases.
- Maximum reduction in resistance occurs when the coatings are relatively thin; increasing the coating thickness further results in progressively diminishing effect. This is because of the additional bulk resistance of the coating material. Indeed, one can see the slight rise in the reduction factor for large values of (t/b) combined with the large (unrealistic) value of $(a/b) = 0.4$
- The presence of the gas slightly reduces the constriction resistance, hence reducing the effectiveness of the coating

In a later work, Olsen et al. (2002) extended the modelling to include radiation. The effect of radiation was similar to including a conducting gas in the gap—the reduction in constriction resistance leading to a reduction in the constriction resistance reduction factor.

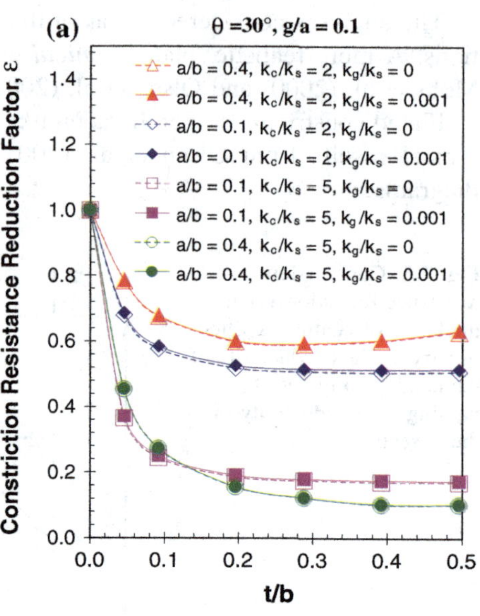

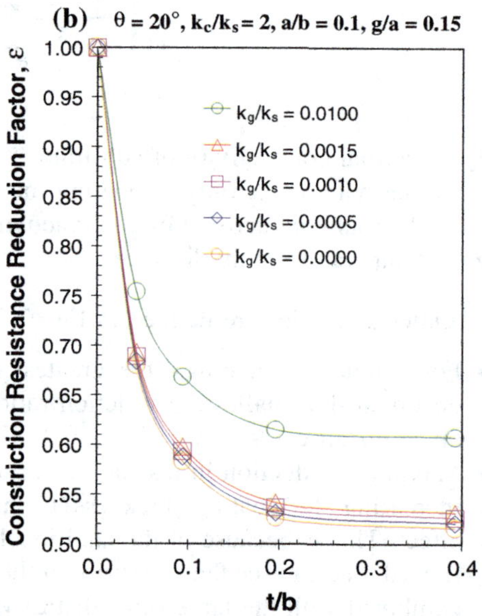

Fig. 7.6 Coated conical constriction—model for numerical analysis

7.3.2 Thermal Contact Conductance of Coated Surfaces

The discussion, so far, refers to the constriction of a single contact spot on a coated surface Analytical and experimental studies also exist of the TCC of the whole coated surface. Some of these are discussed in this section.

7.3.2.1 Effective Hardness

It is, of course, important to realise that the microhardness of coated surfaces will be a composite property that depends on the hardness of both the substrate and the coating material. It may also depend upon the method used for coating the material.

Antonetti and Yovanovich (1985) determined the effective microhardness on the basis of experimental observations on Nickel 200 specimens coated with vapour deposited silver (soft coating layer on a hard substrate). It was noted that, when the coating is thin, the hardness is mainly controlled by the hardness of the substrate. When the coating is thick, however, the hardness of the coating is the controlling factor. *The following relations were deduced from the experimental data and are applicable to only the particular coating (silver on nickel)*:

$$H' = H_s\left(1 - \frac{t}{d}\right) + 1.81H_c\left(\frac{t}{d}\right) \tag{7.7a}$$

$$H' = 1.81H_c - 0.208H_c\left(\frac{t}{d} - 1\right); \quad 1 \leq \left(\frac{t}{d}\right) \leq 4.9 \tag{7.7b}$$

In Eqs. (7.7a and 7.7b), the subscripts s and c refer to the substrate and the coating, respectively; t is the thickness of the coating and d is the equivalent indentation depth of the harder contacting surface obtained from a Vickers microhardness test.

Rajamohan and Madhusudana (1998, 1999) reported on the results of several series of microhardness tests, combined with Finite Element Analysis, on uncoated as well as coated surfaces. The coating method, in all cases, was by electroplating. Both soft and hard coatings were considered. The finite element modelling was first validated by comparing the results with published and measured microhardness data. Only some representative results are shown in Fig. 7.7.

The following general conclusions could be drawn from the results shown:

- For bare metals, the hardness varied with the indentation depth; an indentation size effect (ISE) was present for all metals.
- For hard coatings, e.g., nickel coating on mild steel, after the rapid initial decrease from a high value, the effective hardness decreased gradually with the indentation depth and eventually approached the hardness of the substrate material.
- For soft coatings, two distinct regimes can be identified. The first part consists of an initial high value for the hardness due to ISE rapidly decreasing to approach the hardness of the coating material. (This hardness is indeed reached, if the material is soft and the coating is thick.) In the second part, the effective hardness increased after reaching the minimum value and eventually approached the hardness of the substrate. This latter event occurred when the indentation depth was about five times the coating thickness.

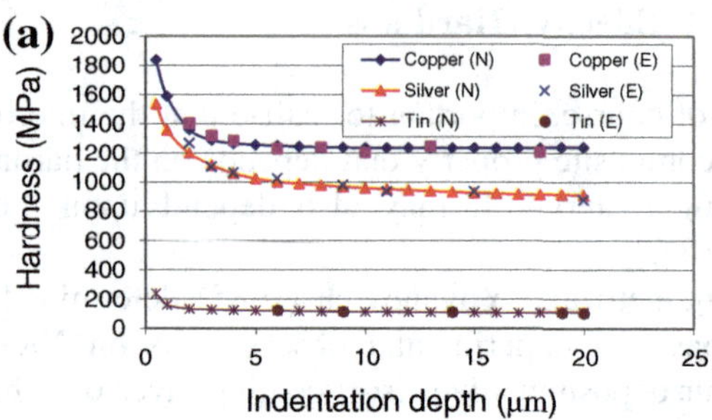

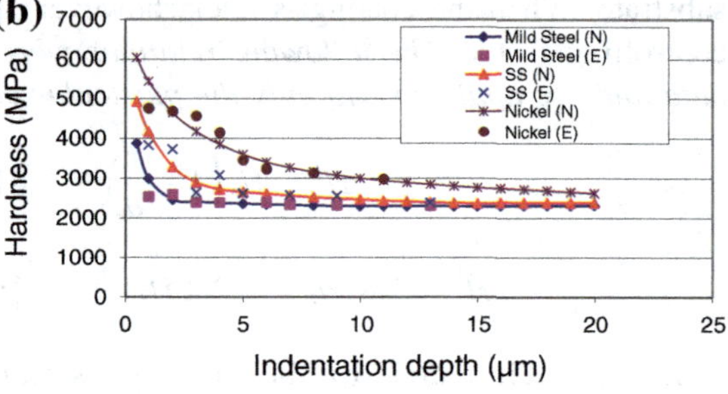

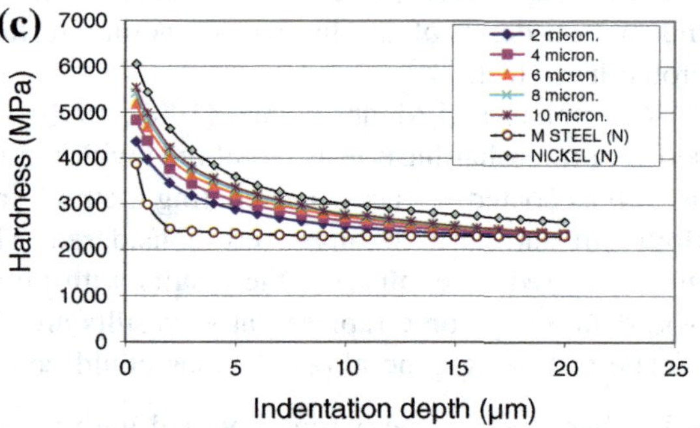

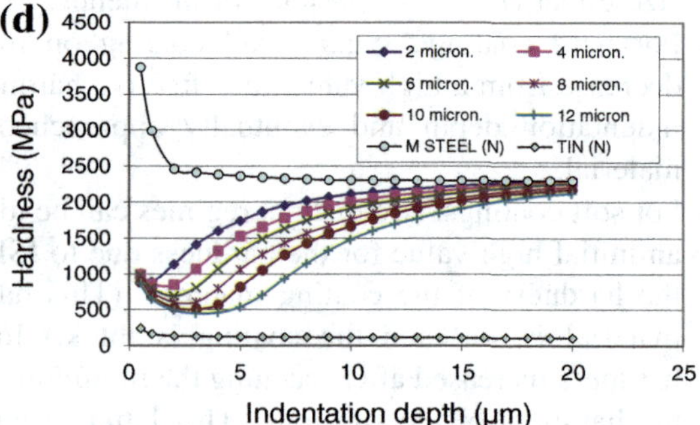

Fig. 7.7 **a** Variation of microhardness vs indentation depth for "soft" metals (N = Numerical; E = Experimental), **b** Variation of microhardness vs indentation depth for "hard" metals, **c** Variation of microhardness vs indentation depth for nickel coated mild steel, **d** Variation of microhardness vs indentation depth for tin coated mild steel

7.3.3 Results for TCC of Coated Surfaces

The experimental data of Peterson and Fletcher (1990) on anodised aluminium coatings, varying in thickness from 60.9 to 168.3 μm in contact with an uncoated aluminium 6061-T6 surface yielded the correlation:

$$\left(\frac{ht}{k}\right)\left(\frac{t}{\sigma}\right)^{0.25} = 0.83\left(\frac{P}{H}\right) \times 10^{-2} + 0.11 \times 10^{-4} \quad (7.7c)$$

It is noted that any variation in the surface thermophysical properties, due to anodisation, was not reflected in the above correlation; also only one of the surfaces was anodised.

An experimental study of the effect of metallic coatings on the TCC of turned surfaces was reported by Kang et al. (1990). The following points are noted with reference to these tests:

- The bare metal was aluminium 6061-T6
- The coating materials were tin, indium and lead
- Four different thicknesses of each coating were prepared
- For each coating material there existed an optimum thickness that yielded the maximum conductance
- The optimum thicknesses for tin, indium and lead were 0.2–0.5, 2–3, and 1.5–2.5 μm respectively.

A new (in 1992) type of surface coating, the transitional buffering interface (TBI) was used in the studies of Chung et al. (1993a) and Sheffield and Chung (1992). This process involved plasma enhanced deposition of a thin coating of a two-phase mixture of either, copper and carbon, or silver and carbon. In each case, the relative ratio (Cu:C or Ag:C) could be altered by changing the deposition parameters, thus giving the desired chemical gradient through the coatings. This process is said to offer excellent characteristics of adhesion of the coating to a wide range of base materials as well as a close control of coating thickness and surface roughness. Chung et al. found that the silver-carbon coatings produced a stronger adhesion than pure silver coatings. However, it was also noted that, for both rough and smooth surfaces, the enhancement produced by coatings of silver and copper were significantly higher than those produced by silver-carbon and copper-carbon, respectively. Further, Sheffield and Chung noted that the improvements in conductance of the order of 100 % were observed when both surfaces were coated compared to the case when only one surface was coated. This confirmed the observation of Mikic and Carnasciali (1970), noted earlier in this chapter.

Experimental results for the TCC of a ceramic (Al_2O_3) coated with (typically 0.2–0.3 μm) aluminium, copper and iron carbide have been reported by Chung et al. (1993b). As expected, an enhancement was noted with aluminium and copper coatings, but the TCC was reduced with the iron carbide coating.

Other results for coated surfaces are listed in Table 7.4

Table 7.4 Summary of representative results for TCC of coated surfaces to 1993

Details	Antonetti and Yovanovich (1985)	Peterson and Fletcher (1990)	Marotta et al. (1994)	Lambert and Fletcher (1992)	Lambert and Fletcher (1993)
Substrate material	*Nickel 200*	*Aluminum 6061-T6*	*Aluminum 6101-T6*	*Aluminum A351-T6*[3]	*Aluminum A351-T6*[3]
Conductivity	77	179	207	151	151
Hardness	2750	980	410	980	980
Roughness	a. 1.27 b. 4.27 c. 8.32	1.00 (0.98–1.04)	0.87 (average)	0.45 (average)	0.45 (average)
Coating material	*Silver*	*a. Tin* *b. Indium* *c. Lead*	a. *Titanium Nitride* *b. Beryllium Oxide*	*a. Gold* *b. Silver*	*Silver*
Method of coating	Vapour deposited	Vapour deposited	Physical vapour deposited	Vapour deposited	a. Electroplated b. Flame sprayed
Conductivity	425	a. 65.8 b. 81.7 c. 37.0	a. 24 b.	a. 315 b. 427	427
Hardness	390	a. 49 b. 9.8 c. 39.2	a. 2600 b. 4900	a. 1250 b. 1250	1250
Thickness	a. 1.2; 6.3 b. 0.81; 1.4; 39.5 c. 2.4; 7.2; 18.0	a. 0.25; 0.76; 1.82 b. 0.28; 2.54; 3.71 c. 0.25; 1.8; 5.0	a. 3 b. 3	a. 1; 3 b. 1; 3	a. 13; 51 b. 13; 76

(continued)

Table 7.4 (continued)

Details	Antonetti and Yovanovich (1985)	Peterson and Fletcher (1990)	Marotta et al. (1994)	Lambert and Fletcher (1992)	Lambert and Fletcher (1993)
Pressure	*500–3700 k Pa*	*75–1600 k Pa*	*170–700 k* Pa[4]	*170–860 k Pa*	*170–860 k Pa*
$h_{\text{bare, vac}}$	a. 2600–12000 b. 1100–8000	1000–10500[2]	3500–8800	700–2000	700–2000
h_{coated}	a. 12000–58000 20000–70000[1]	a. 170–12800	a. 2200–7900	a. 1300–3000; 600–1800;	a. 1900–4500; 800–2100
	b. 2200–12000 4000–23000 11000–95000	b. 1920–33370	b. 3000–9200	b. 550–2500; 1200–4000	b. 420–1250; 400–900
	c. 2200–14000 5200–33000 9500–58000	c. 1990–22460			
h_c/h_{bv}	a. 4.61–4.83; 7.69–8.75	a. 1.7–1.22; 0.85–0.75; 0.35–0.50	a. 0.65–0.90	a. 1.86–1.5; 0.86–0.9	a. 2.7–2.25; 1.15–1.05
	b. 2–1.5; 3.64–2.88; 10–11.88	b. 2.6–1.9; 7.0–2.8; 2.3–1.8;	b. 0.85–1.05	b. 0.78–1.25; 1.70–2.00	b. 0.6–0.62; 0.57–0.45
	c. 2.2–2.59; 5.2–6.11; 9.5–10.74	c. 1.5–1.2; 4.2–2.2; 2.0–1.2			

[1] At 1800 kPa; the corresponding bare joint conductance was 8000 W/m^2 K
[2] Average for 12 pairs of surfaces
[3] The other surface, common to all contacts, was electroless nickel-plated copper
[4] Range for which the enhancement is reported

7.3.4 TCC of Coated Surfaces: Recent Works

Li et al. (2000) conducted an extensive investigation of TCC of mild steel (plain carbon steel) surfaces electroplated with tin (eight specimens), silver (nine specimens) and copper (eight specimens). Nine stainless steel specimens were coated with aluminium using a filtered arc deposition (FAD) process. FAD is a vapour deposition process in which a magnetic filter is used to filter macroparticles from being deposited on to the target surface. The thermophysical properties of the substrate and coatings are summarised in Table 7.5.

Each of the contacting surfaces had been mechanically polished and then bead blasted, to obtain a flat uniform rough surface, prior to coating. In each case, it was noted that the surface characteristics altered slightly after coating. In general, coating reduced the surface roughness and slope and increased the radius of curvature. The exception was with copper plating where the trends were opposite. All of the heat transfer tests were conducted in vacuum.

Test results are shown below for silver plating on mild steel and aluminium coating on stainless steel. The results for tin and copper platings showed similar trends. All of the results are reported in Li et al. (2000) (Fig. 7.8).

It is clear from these tests that an optimum thickness of coating existed in all cases although its value could not be determined exactly, but it appeared to be nearer to 10 μm than 20 μm in all of the tests. In fact, at 20 μm thickness of silver coating and at low contact pressures, the conductance is back to the figure for uncoated surfaces. This means that, at these low pressures, whatever gains were made by improving the thermal characteristics of the surfaces were cancelled by the increased bulk resistance of the coating material. We observe similar trends for the aluminium coated stainless steel specimen. Of course, this is logical when we observe the law of diminishing returns as the coating thickness is increased in the numerical results for coated constrictions (Fig. 7.6). The optimum thickness values are somewhat higher than those reported by Kang et al. (1990), but lower than those anticipated by Kharitonov et al. (1974).

Joints used for cryogenic structural supports and satellite deployment mechanisms are often exposed to very low temperatures. A conducting or insulating interstitial medium is introduced at the interface for controlling the contact conductance. A reduction in contact conductance is achieved by providing a non-metallic coating at the interfaces. Molybdenum sulphide (MoS_2) is widely used as

Table 7.5 Properties of coating materials and the non-dimensional parameter combining substrate and coating material properties

Coating material	k_c (W/(mK)	H_c (MPa)	k_c/H_c	$H_s\, k_c/(H_c\, k_s)$
Tin	67	103	0.65	34.65
Silver	427	1055	0.40	21.20
Copper	398	1230	0.32	16.96
Aluminium	237	1277	0.19	45.12

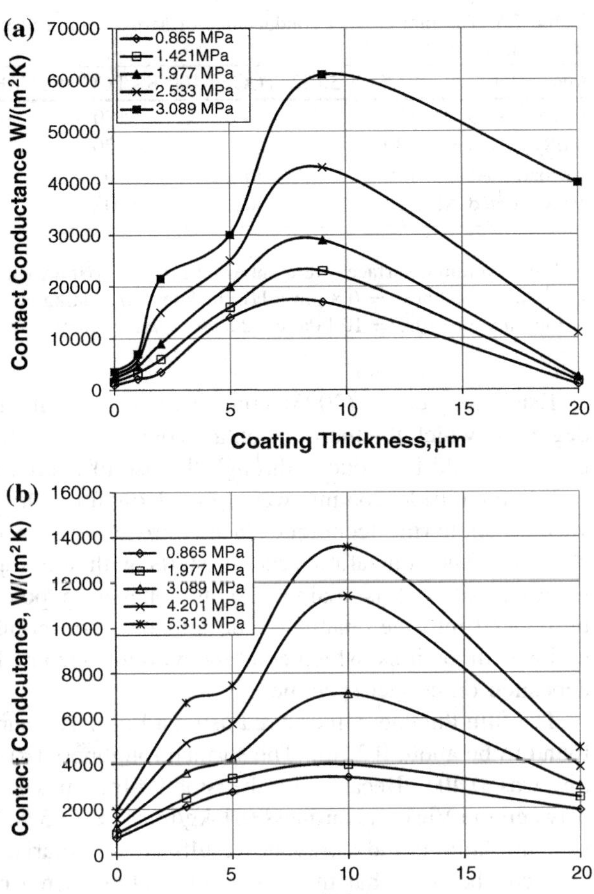

Fig. 7.8 **a** Mild steel specimens with silver plating; zero thickness means uncoated surface. **b** Stainless steel specimens with aluminium FAD coating; zero thickness means uncoated surface

the nonmetallic coating in cryogenic structural joints, because it not only reduces the joint conductance but also provides dry lubrication at the joint. Ramamurthi et al. (2007) tested eleven pairs of specimens made of aluminum (Al) and stainless steel (SS). Each sample pair was 25 mm in diameter and either 25 or 5 mm in height, respectively. They were prepared with varying levels of surface finish, with values of CLA roughness between 0.2 and 2 μm and the average slope of asperities varying from 0.09 to 0.15. The measured flatness values were well within 0.2 μm.

One side of each of the aluminum samples was anodized to prevent atmospheric oxidation of the surface and to obtain better retentivity of the MoS_2 coating on the surface.

Tests were done at temperatures between 50 and 300 K. Typical results as shown in Table 7.6 indicated that MoS_2 coating over aluminium and stainless steel surfaces significantly reduced the thermal contact conductance at cryogenic temperatures.

Table 7.6 Thermal contact conductance of Molybdenum disulphide coated surfaces (results of Ramamurthi et al. 2007)

Material ↓	TCC at 50 K	TCC at 100 K	TCC at 150 K	TCC at 200 K	TCC at 250 K
Uncoated SS	160		180	250	500
MoS_2 coated SS	30		70	100	200
Uncoated Al	180	270	400	600	860
MoS_2 coated Al	90	130	205	330	530

Note

1. The aluminium surfaces were anodised prior to coating with MoS_2
2. Surface roughness = 0.8 μm (cla). thickness of coating = 5 μm
3. Interface Pressure = 10 kPa; conductance values in W/(m^2 K)

Kshirsagar et al. (2003) conducted an experimental study to determine the degree by which the thermal contact conductance at the interface of OFHC copper contacts could be reduced through the use of sputtered silicon nitride films.

Silicon nitride coating was applied on four OFHC copper specimens using vacuum sputtering technique. In this process, argon is admitted into the vacuum chamber. The ionization process starts with the supply of voltage. The target material (silicon) is maintained at a negative potential. Nitrogen gas is then introduced into the chamber. The ions of argon bombard the target material and remove silicon ions, which combine with nitrogen and form silicon nitride and get deposited on copper specimen.

The film thickness, measured using a Form Talysurf (Taylor Hobson, UK), was found to be about 0.2 μm. The surface roughness (σ) was 0.211 μm and the slope (m) was 0.404. Because of the small film thickness, there was no difference between the Vickers Hardness (81 kg/mm^2 $\approx$ 795 MPa). Tests were conducted in vacuum, helium and nitrogen. Results are summarised in Table 6.7.

It can be seen that there is a substantial decrease in conductance due to the coating by silicon nitride in each case. The reduction in helium is particularly noticeable. As noted in Chap. 4, the increase in conductance (in incoated specimens) due to the presence of a gas is not proportional to the gas thermal conductivity (Table 7.7).

Table 7.7 Decrease of thermal contact conductance due to silicon nitride coating on OFHC copper (based on the results of Kshirsagar et al. (2003)

Contact pressure range (MPa)		2–12
Thermal contact conductance range [kW/(m^2 K)]	Uncoated in vacuum	30–95
	Coated in vacuum	7–21
	Uncoated in nitrogen	42–103
	Coated in nitrogen	15–24
	Uncoated in helium	52–126
	Coated in helium	22–30

7.4 Insulating Interstitial Materials

In applications such as isolation of spacecraft equipment, cryogenic storage tanks and, in general, wherever strong insulating supports are required it is necessary to increase the TCR.

The use of silicon carbide, molybdenum disulphide and iron carbide coatings to increase the TCR has already been noted in the previous section.

The use of wire screens as thermal isolation materials has been reported by several investigators. Fried and Costello (1962), and Al-Astrabadi et al. (1977) used copper wire meshes while Gyorog (1971), Sauer et al. (1971a, b), O'Callaghan et al. (1975) used stainless steel wire screens. The method of weaving the screens results in the weft being a series of almost straight wires, all in one plane with the warp interlaced. Contact, therefore, occurs only between the warp and the solid surface, that is at *every other* crossing. The theory of Cividino and Yovanovich (1975) which assumed that the contact occurred at every crossing, therefore, overestimated the conductance. If large scale surface irregularities are present, the screens may reduce the (macroscopic) resistance by providing more heat flow paths in the interface. Thus the presence of wire screens might increase or decrease the interface resistance depending on whether the surfaces are flat and conforming or have waviness and flatness deviations. It is to be further noted that any such change in resistance would also depend upon the parent material/screen material combination.Other insulating materials which have been considered in the past include felt, mica, Teflon, carbon-black filled elastomer and neoprene. The results for these have been summarised in Table 7.8.

7.5 Lubricant Films and Greases

It is to be expected that, when the interface is charged with a grease, the conductance would be noticeably increased due to establishment of better thermal bridges across the interstitital gaps. When the use of thermal greases for enhancement of conductance is considered, the following observations are worth noting (Gwinn and Webb 2003).

Thermal Greases are composed of a thermally conductive filler dispersed in silicone or hydrocarbon oil to form a paste. "Printable" greases can be screen-printed on to a heat sink base plate at a specified thickness.

Thermal greases provide high thermal performance at small contact pressures. They have the ability to fill the interstices with a material whose thermal conductivity is much higher than that of air. For example the conductivity of ShinEtsu G751 thermal grease is 4.5 W/(m K) which is 172 times higher than that of air at room temperature. They do not require curing. Some microprocessor manufacturers recommend the use of thermal grease.

Table 7.8 Thermal contact conductance with insulating interstitial materials

Details	Fletcher et al. (1969)	Gyorog (1971)	Fletcher and Miller (1973)
Substrate material	*Aluminum 2024-T4*	*Stainless steel AISI304*	*Aluminum 2024-T4*
Conductivity	210	15	210
Hardness		2550	
Roughness	0.076–0.152	0.076–0.152	0.381
Interstitial material	*a. WRP-AX-AQ Felt* *b. Mica* *c. Teflon*	*a. WRP-AX-AQ Felt* *b. Mica* *c. Teflon*	*a. Carbon black filled fluorocarbon elastomer* *b. Neoprene* *c. Silver coated copper powder filled silicone elastomer*
Conductivity	a. 0.069 b. 0.363 c. 2.34	a. 0.069 b. 0.363 c. 2.34	
Density, kg/m^3	a. 288 b. 208 c. 160	a. 288 b. 208 c.1 60	a. 1875 b. 645 c. 3650
Thickness, mm	a. 4.47 b. 0.05 c. 0.05	a. 4.65 b. 0.076 c. 1.57	a. 2.1 b. 2.9 c. 0.71
Pressure	*690–2070*	*690–2070*	*690–2070*
$h_{\text{bare, vac}}$	7100–15000	360–2240	4300–9600
h_{insul}	a. 5–9 b. 280–730 c. 1730–2310	a. 3–5 b. 150–300 c. 16.5–17.5	a. 90–135 b. 125–190 c. 1470–1730
$\frac{h_{\text{insul}}}{h_{\text{bare, vac}}}$	a. 0.0007–0.0006 b. 0.039–0.049 c. 0.244–0.154	a. 0.008–0.002 b. 0.217–0.134 c. 0.046–0.008	a. 0.021–0.019 b. 0.029–0.028 c. 0.342–0.251

On the other hand, application of greases may be messy to apply and difficult to remove. Excess grease that flows out of the joint must be removed to prevent contamination and possible electrical short circuits. Grease joints can dry out with time resulting in increased thermal resistance. Grease joints can degrade with higher temperatures and thermal cycles and their thermal performance may deteriorate with time.

Cunnington (1964) conducted tests in vacuum at contact pressures ranging from 40 to 80 psi (276–551 kPa) and observed that the conductance of an aluminium joint increased by more than an order of magnitude when contact surfaces were coated with DC-340 paste (a heat sink compound). In fact, the improvement in conductance was much greater than that obtained by the insertion of indium foil, which is one of the most effective ways for enhancing the TCC. The use of a silicone grease was less effective although it produced enhancements of similar magnitudes to that obtained with indium foil. The reason was that the conductivity of silicone grease was estimated to be only half that of DC-340. It was further noticed that the results for greases and pastes would be applicable to a wide range of joint configurations because of their ability to flow and fill the interstitial spaces at relatively low contact pressures. In other words, grease-filled joints would be less sensitive to changes in roughness and flatness of surfaces.

Experimental results for stainless steel surfaces treated with four different types of lubricant films, namely, silicone spray, Molykote, lithium and graphite greases, have been reported by Sauer et al. (1971a, b) and Sauer (1992). All four lubricants provided enhancement, 8 to 70 fold when compared to vacuum conductance and 0–60 % when compared to conductance in air of the bare joint. It was noted that lithium grease was the most effective and the graphite the least. The mean junction temperature in the tests was about 90 °C.

7.6 Other Interstitial Materials

7.6.1 Phase Change Materials

In electronic packaging, heat generating components are typically mounted on printed circuit boards, which must be removed from the housing or rack for periodic maintenance. Thus the heat generated must pass through multiple mechanical interfaces before being absorbed by a remote cooling medium. A novel concept for reducing the TCR, using a low melting point interface material, was first proposed by Cook et al. (1982). At room temperature, the alloy is in solid state to facilitate ease of handling during assembly and disassembly. During operation, heat transferred to the joint raises the interface temperature causing the alloy to change phase from solid to liquid. Thus a continuous metallic heat transfer path is provided, significantly reducing the TCR. The authors proposed three ways in which the low melting point alloy may be inserted between the contacting surfaces:

1. As a thin sheet by itself
2. As a filler in a thin porous metal structure
3. As coatings on both sides of a thin solid metal sheet or film.

The low melting point alloy used by Cook et al. is an example of true phase change material (PCM). The inorganic salts used in thermal energy storage are also genuine PCM's where the latent heat of solidification, is the key factor. In current literature on elctronics cooling, however, PCMs are assumed to mean a mixture of suspended particles of high thermal conductivity such as fine particles of a metal oxide and a base material (Gwinn and Webb 2003). These materials do not actually change phase, but their viscosity diminishes so that they flow. The base material can be refined paraffin, a polymer, a co-polymer or a combination of all of these. The base material is solid at low temperatures, but behaves much like grease after reaching the "phase change" (typically 50–90 °C).

PCM has the ability to flow throughout the thermal joint to fill the air gaps and provide minimum thickness. When the joint becomes thin, the viscosity prevents 'pump out' from the interface. It handles easily, like a thermal pad, for installation. A compressive force is needed, however, to bring the surfaces together and cause the TIM to flow.

In relation to thermal management of electronic packages, Ochterbeck et al. (1990) noted that the enhancement of the TCC while simultaneously insulating electrically could be quite challenging. Out of the materials they tested, only Isostrate, which is a Kapton MT coating (a DuPont polyimide film) could provide an increase in the TCC while all other electrically insulating materials caused a decrease in the TCC.

Note Recent work on othe interstitial materials can be found on a later section in this chapter. This is because they need an understanding of the characteristics of carbon nanotubes which are discussed below.

7.7 Carbon Nanotubes

Graphene, a single layer of graphite, is a one atom thick sheet of carbon arranged hexagonally. Carbon nanotubes are graphene layers rolled into a cylindrical shape.

Carbon nanotubes (CNT's) and carbon nanofibres (CNF's) possess excellent thermal conductivity along their length due to their ability to transmit heat by ballistic conduction, that is, the phonons are free to move without being scattered by impurities or defects. Note that the phonon mean free path, about 1.5 nm, is of similar order of magnitude as the diameter of the nanotube. But their conductivity in the lateral (radial) direction is poor. Room temperature thermal conductivity of a single wall carbon nanotube (SWNT) is estimated to be 3500 W/(mK) along its axis (Pop et al. 2006). By contrast, the thermal conductivity across the axis, of a SWNT is about 1.52 W/(mK), Sinha and Yeow (2005).

CNT arrays can be effective in reducing thermal interface resistance, potentially satisfying the increasing power dissipation challenge in microelectronics (Fig. 7.9).

Yu et al. (2006) grew CNFs using a plasma enhanced chemical vapour deposition (PECVD). Silicon substrates with a pre-deposited 30-nm-thick Ti barrier layer and a 30-nm-thick Ni catalyst layer were subjected to a glow discharge at a dc bias of 585 V, 500 W, and 0.85 A under a total flow of 100 standard cubic centimetre per minute (sccm) of 4:1 NH_3:C_2H_2 process gas mixture at 4 Torr for

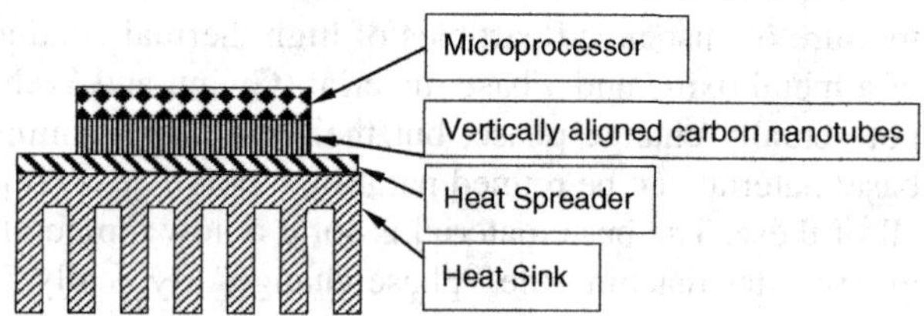

Fig. 7.9 Carbon nanotubes as interstitial material in electronics cooling

45 min. CNF growth rate under these conditions was approximately 500 nm/min. Cross-sectional transmission electron micrographs were obtained to investigate the CNF quality and graphitic microstructure.

Yu et al. found that, although the axial thermal conductivity of individual multi-walled carbon nanotubes is high, the effective thermal conductivity of CNT mats and CNT bundles was one or two orders of magnitude lower than that of individual defect-free CNTs due to the large thermal contact resistance between adjacent CNTs in the bundles. The contact thermal resistance, at the nanometre scale, of point and line contacts between a CNT or CNF and a planar surface can also be high due to enhanced phonon-boundary scattering at the nanocontacts and should not be ignored.

Arrays of vertically aligned CNTs can potentially take advantage of the high axial k of individual tubes but are limited in part by the low packing fraction of CNTs in the array and the interface resistance at the CNT–substrate contacts. A simple model of the aligned CNT arrays as conductors in parallel suggests that k should scale with the packing fraction. However, results of several experimenters show that the effective k is lower than predicted by the volume fraction indicating that factors beyond the low density of vertically aligned CNT arrays, such as interface resistances between the CNT array and surrounding materials, degrade thermal performance. Gao et al. (2010) studied the unique implementation of CNTarray interface materials grown directly on a thermoelectric material. In particular, they characterized the thermal properties of 1.5-μm-thick metal-coated aligned multiwalled carbon nanotube (MWCNT) films on SiGe, a standard high-temperature thermoelectric material, using a nanosecond thermo reflectance technique. They found that the thermal interfaceresistances between the CNT film and surrounding materials were the dominant barriers to thermal transport, ranging from 1.4 to 4.3 (m^2 K)/MW. The volumetric heat capacity of the CNT film was estimated to be 87 kJ/(m^3 K), corresponding to a volumetric fill fraction of 9 %. They suggested that the boundary resistances can be reduced in future studies by the application of a eutectic-based binder material that maintains good thermal contact with more tubes and by improving the CNT growth.

Son et al. (2008) investigated the interface thermal resistance (ITR) of the native interface between vertically aligned multi-walled carbon nanotube arrays and the SiO_2/Si substrate.

Three vertically aligned MWCNT specimens with volume fraction of $\approx$2 % were grown on SiO_2/Si substrates using a chemical vapour deposition (CVD) method. A mixture of xylene (as a carbon source) and ferrocene (as an iron catalyst source) in gas phase (1 g/100 ml), preheated up to 200 °C in the flexible heater, was fed into the tube furnace at 770 °C at a rate of 0.11 ml/min in an Ar/H_2 atmosphere. Finally, three specimens, 128, 215, and 460 μm lengths of the CNTs in the film form, were prepared by controlling the CVD growth time. In all specimens, the averaged outer diameter (D_{outer}) and inner diameter (D_{inner}) of the CNTs were $\approx$30 and $\approx$10 nm, respectively. The distance between CNTs of $\approx$188 nm was deduced from the volume fraction (Fig. 7.10).

To calculate the ITR, they adopted an analytical model proposed by Prasher et al. in which the ITR effect across a nanoscale contact is obtained by summation

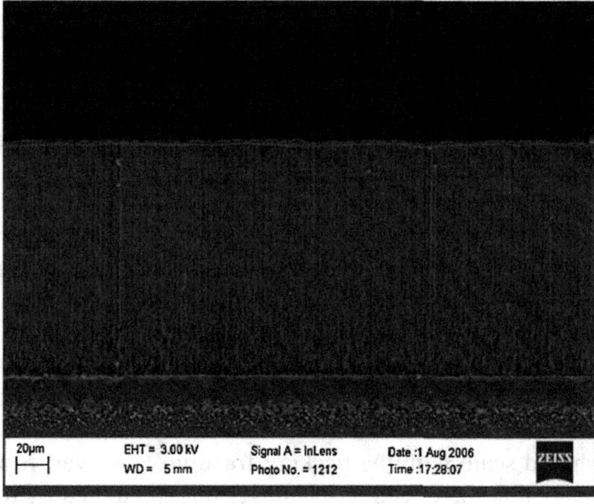

Fig. 7.10 Scanning electron microscopy (SEM) image of the 128 μm thick MWCNT array grown on the SiO_2/Si substrate (Son et al. 2008. Reprinted by permission from American Institute of Physics)

of the macroscopic constriction resistance (CR) from classical heat diffusion and the microscopic thermal boundary resistance (TBR) due to scattering and transmission of the microscopic heat carriers at the interface. Based on our estimations using the kinetic theory, the mean-free path ($\approx$20 nm) of heat carriers in the CNT material under investigation is comparable to the characteristic length scale ($\approx$30 nm); hence the model could be used.

With this modelling, the ITR was estimated to be $\approx 4.4 \times 10^{-7}$ m^2 K/W (where CR $\approx 4.0 \times 10^{-7}$ and TBR $\approx 4.4 \times 10^{-8}$ m^2 K/W).

The *measured* specific ITRs of the native interface between the CNT film and substrate were determined to be 4.9×10^{-5} and 4.8×10^{-5} m^2 K/W for samples of length 128 and 215 μm respectively,

Note that the theoretical values are two orders of magnitude lower than the experimental values. This implies that imperfect contact may have a vital role in the heat flow at the interface causing the discrepancy between the experimental and theoretical results. In addition to imperfect contact, the existence of catalyst particles in between the CNT and the substrate provides another interface where consecutive phonon-electron–phonon scattering occurs.

Xu (2006) made an extensive study of array of carbon nanotubes as an Si-supported CNT array interfaced with Cu, that is, the CNT array was sandwiched between copper and silicon substrates. Multiwalled carbon nanotube arrays were synthesised on various substrates by plasma enhanced chemical vapour deposition method, Dense (10^8–10^9 CNTs/mm^2) and tall (10–100 μm) CNT arrays were created over areas of the order 1–10 cm^2.

He noted that (refer to Fig. 7.11):

- some CNT's buckled
- the free tips contacted the copper surface at different angles
- some CNT's did not make direct contact with the copper surface

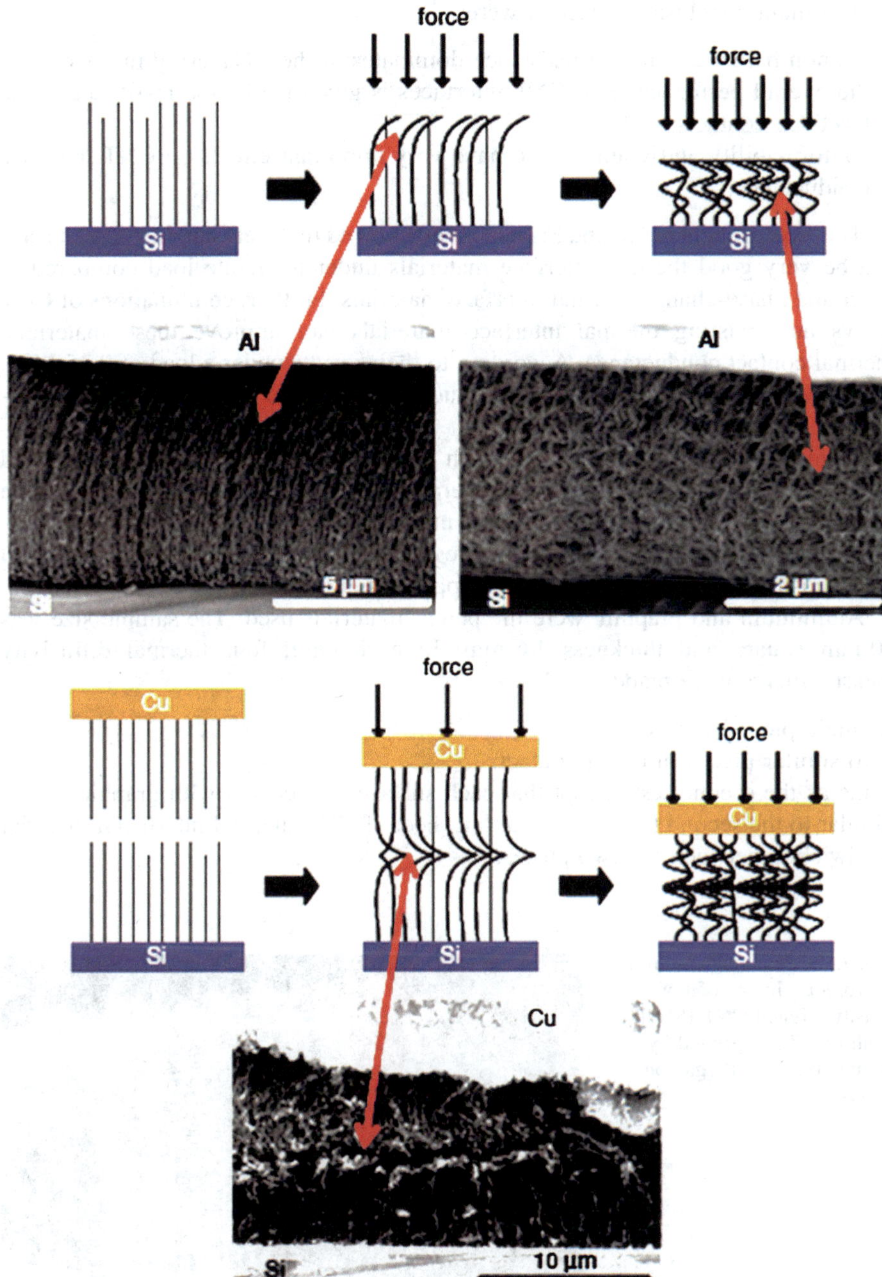

Fig. 7.11 SEM images of a substrate-supported CNT array at various degrees of compression. The CNTs have an average diameter of approximately 20 nm, and a volume ratio of approximately 15 %. Illustrations and SEM image (*below*) of two CNT arrays in contact, show the bending of CNTs at the interface (Cola et al. 2009. Reprinted from permission by Pergamon)

The main conclusions reached were:

- phonon ballistic transport resistance dominates at the CNT array interface
- the overall performance of CNT interfaces is governed by the resistance at the CNT-Cu contacts, and
- conformability and volume ratio have very important effects on CNT interface conductance.

The experiments of Xu and Fisher (2006) showed that free-standing CNT arrays can be very good thermal interface materials under moderate load compared to sheet and phase-change thermal interface materials. Further, combinations of CNT arrays and existing thermal interface materials can improve these materials' thermal contact conductance. According to their results under a load of 0.35 MPa, the PCM–CNT array combination produced a minimum thermal interface resistance of 5.2 mm^2 K/W.

Using quartz as the substrate, Shaikh et al. (2007) managed to grow aligned CNTs with thickness ranging from several micrometers to about 200 μm with a narrow diameter distribution around 15 nm (Fig. 7.12).

The thermal diffusivities of the samples, with and without the interface material was measured in a modern light flash apparatus (LFA 447).

Aluminium and graphite were the parent materials used. The sample size was 10 mm square and thickness 1.6 mm. In each case, four thermal diffusivity measurements were made:

A single piece by itself.
Two similar pieces in direct contact.
Same as the second test except that each surface was coated with graphite.
Similar to the second test but with a thin sheet of CNT thermal interfacial material sandwiched between the two pieces.

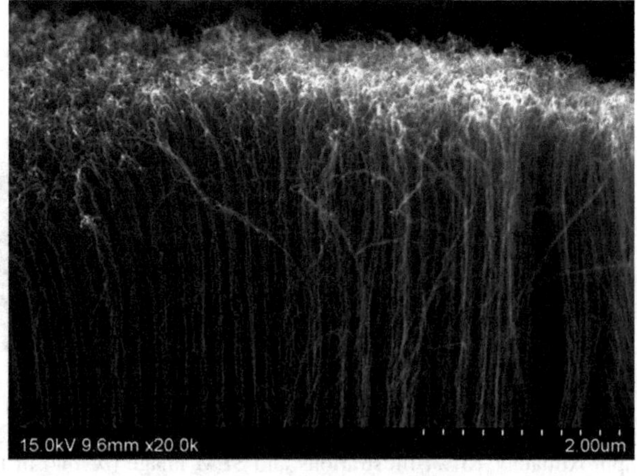

Fig. 7.12 Scanning electron microscopy image of low density aligned CNT (Shaikh et al. (2007). Reprinted by permission from Pergamon Publishers)

In each case, the specific heat capacity could also be measured in the same device. Measuring the density of the materials separately, the corresponding thermal conductivity can be calculated (Fig. 7.13).

The following results were obtained as a result of their experiments.

Sample type	Thermal resistance (K/W)
Aluminum pieces in direct contact	3.573
Aluminum pieces with graphite interface	0.842
Aluminum pieces with CNT interface	0.736
Graphite pieces in direct contact	2.523
Graphite pieces with graphite interface	0.679
Graphite pieces with CNT interface	0.562

Note Since the sample area was 100 mm^2 , these figures need to be multipled by 100 so that the results will be in (mm^2 K/W) and comparison can be made with other results

Desai (2006) noted that, in practice, the nanotubes are grown off a surface such as silicon and the height to which the nanotubes grow cannot be controlled to great precision. Hence there will be a small gap between some of the nanotubes and the aluminum interface. His results indicated that, despite the effects of height variation, a thermal interface material with vertically aligned carbon nanotubes has the potential to be a high thermal conductivity thermal interface material.

Cola et al. (2009) noted that in applications where the materials that form the interface cannot be exposed to the temperatures normally required for CNT growth, direct synthesis of CNT array interfaces may be difficult. This is particularly true when interface surfaces are relatively rough (e.g., unpolished Cu–Cu interfaces), requiring CNT arrays that are dense and long enough to fill the interface voids effectively.

Cola et al. fabricated a CNT thermal interface material (TIM) that consists of CNT arrays directly and simultaneously synthesized on both sides of a copper foil has been fabricated. This TIM eliminated the need for exposing temperature-sensitive materials and devices to normal CNT growth conditions and provided greater conformability to rough interfaces due to foil deformation that increases the number density of contact points between free CNT tips and their opposing substrate. The CNT/foil TIM is similar to existing state-of-the-art TIMs in that it

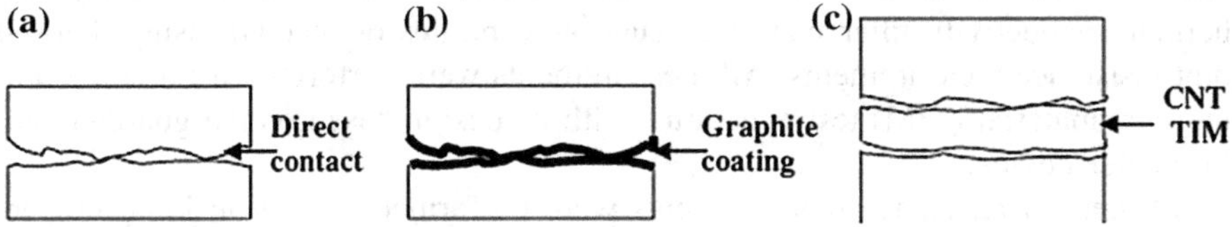

Fig. 7.13 Schematic of contacting surfaces: **a** in direct contact; **b** separated by graphite coating; **c** separated thin sheet of CNT interface material (Shaikh et al. 2007. Reprinted by permission from Pergamon Publishers)

can be inserted into several different interface configurations; it is also dry, removable, and has an intrinsically high thermal conductivity.

It was noted above that Xu and Fisher considered that combinations of CNTs and existing interfacial materials could improve the TCC of these materials. On similar lines, Biercuk et al. (2002) used single-walled nanotubes (SWNTs) to *augment* the thermal transport properties of industrial epoxy. They found that epoxy loaded with 1 wt% unpurified SWNTs exhibited a 70 % increase in thermal conductivity at 40 K and 125 % at room temperature.

7.7.1 Carbon Nanotubes: Summary

Although single carbon nanotubes possess high theoretical thermal conductivity along their axis, their effective thermal conductivity in an array is diminished for the following reasons:

- The thermal contact resistance between adjacent tubes in the bundle
- The significant boundary thermal resistance, at nanometric level, at point and line contacts between the tubes and the adjacent plane surfaces
- Some tubes contacting the surfaces at an angle other than 90° and not contacting at all
- Buckling of the tubes.

The performance of carbon nanotube arrays may be enhanced by

- Application of eutectic-based binder material
- Increasing the volumetric fill fraction

The performance of existing materials such as industrial expoxies may be improved by combining them with CNT arrays.

7.8 Other Interstitial Materials: Recent Works

Teertstra (2007) determined the bulk thermal conductivity and thermal contact resistance at the bonded surfaces for thermal adhesive materials (containing high thermal conductivity filler materials, such as ceramics or metals), using thermal joint resistance measurements. All measurements were performed using a thermal interface material (TIM) test apparatus, with its design based on the guarded heat flux meter device.

All thermal resistance measurements were performed at a mean joint temperature of 50 °C with a load of 150 N applied to the test column. Vacuum conditions were maintained (Fig. 7.14).

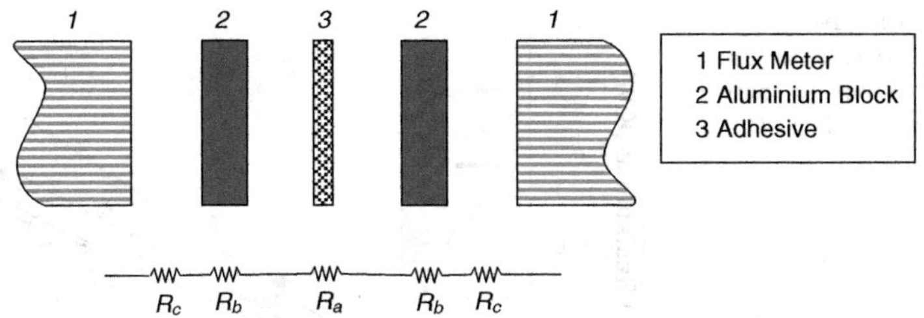

Fig. 7.14 The thermal resistance network considered by Teertstra 2007. Flux meters: Electrolytic iron, Blocks: Aluminium—2024 T-351; 25 × 25 × 6.43 mm

The thermal resistance of an interface can be decreased by the use of a thermal interface material (TIM). The thickness of the TIM is often referred to as the bond line thickness.

The bond line thickness (*BLT*) was calculated by subtracting the thickness of each of the blocks measured at the centre using a micrometer from the total thickness of the test sample after the adhesive had cured.

Preliminary tests with *no adhesive* yielded the following results:

With one aluminium block:

$$2R_c + R_b = 0.204 \quad \text{(a)}$$

With two aluminium blocks:

$$3\,R_c + 2R_b = 0.347 \quad \text{(b)}$$

Comment: Eq. (b) *assumes that the contact resistance between the aluminium blocks and the electrolytic iron flux meters is the same as that between the two aluminium blocks.*

Upon solving (a) and (b) simultaneously:

$R_c = 0.61$ and $R_b = 0.82$ (K/W)

In the actual tests (as shown in the diagram above), the adhesive resistance was given by:

$$R_a = R_j - 2R_b - 2R_c$$

Tests were carried out for four different BLT's for each of the five adhesives. When Ra was plotted against BLT and a straight line fitted in each case. see Fig. 7.15 below, such that:

$$R_a = R_{cont} + \mathrm{k} \times \mathrm{BLT}$$

where R_{cont} the contact resistance is the intercept on the y-axis and k is the thermal conductivity of the adhesive. The results are shown in the table below Fig. 7.15:

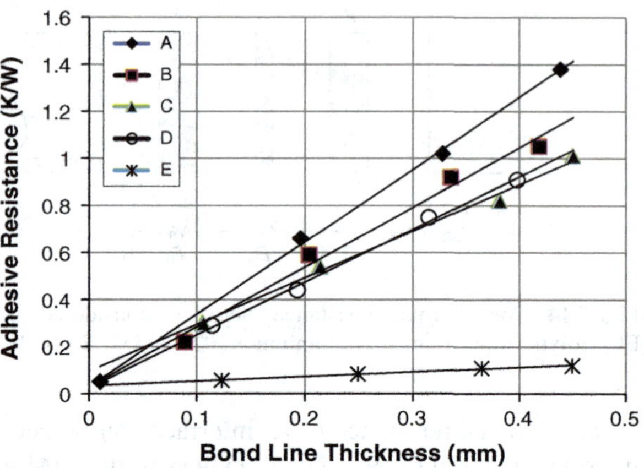

Fig. 7.15 Adhesive resistance versus bond line thickness (composite chart re plotted from the data of Teertstra)

Material	k [(W/(mK)]	R'_{cont} (K/W)	R_{cont} (mm^2K/W)
A High thermal conductivity RTV silicone	0.53	0.013	8.125
B Aluminum-filled 2-part epoxy putty	0.65	0.017	10.625
C Aluminum-filled 2-part epoxy bonding resin	0.84	0.051	31.875
D Silver-filled 2-part epoxy	0.73	0.012	7.5
E Silver-filled thermoplastic	7.8	0.017	10.625

Note

1. The specific thermal resistance values in the last column have been obtained by multiplying the R_{cont} values by the cross-sectional area, 625 mm^2
2. Although the bulk conductivities may be different, the R_{cont} values are not much different. It was postulated that the higher contact resistance of C may be due to the liquid hardener which caused the adhesive to have a lower viscosity
3. At first glance, the specific resistance might appear lower than those obtained with CNT arrays as interface material. It is important to observe that the bare junction here is made of a pair of high conductivity aluminium bars which will have a lower thermal resistance compared to a junction made of say, silicon, surfaces
4. The graph clearly indicates that to minimise contact resistance, the adhesive materials must be applied a thinly as practicable

Abadi and Chung (2011) noted that Thermal pastes with high conformability but low thermal conductivity can perform as well as (or even better than) those of high thermal conductivity but less conforming. The TIM needs to be conformable in order to be able to displace air from the interface by filling the microscopic valleys in the topography of the mating surfaces.

Since the performance of the TIM depends on the structure, especially the roughness, of the proximate surfaces it is important to specify the characteristics of the mating surfaces while describing the performance of the TIM. The thermal conductivity is important but it is not the only factor affecting its performance.

The TIM used in the authors' experiments and numerical modelling was a core sheet coated with polyol-ester based carbon paste.

[A polyol (also known as polyhydric alcohol) is an alcohol having numerous hydroxyl groups. Polyols include polyethers, glycols, polyglycols, polyesters and polyglycerols. They are low molecular weight water-soluble polymers and oligomers. Polyols are classified as diols, triols and tetrols, depending on the number of OH end groups. A diol is linear, with an OH group at each end. Triols and tetrols are branched, such that there is an OH group at the end of each of three and four segments, respectively, that emanate from a point. Polyols constitute a class of organic materials that vary substantially in molecular shape, molecular length and melting temperature, thus providing choices that can suit the requirements of phase-change thermal interface materials].

The core sheets considered were copper, aluminium, indium and flexible graphite. Flexible graphite is advantageous in its resiliency and low modulus, while the metals are advantageous in their high thermal conductivity and small thickness. Among the three metals, indium has the lowest modulus and the lowest thermal conductivity. The schematic representation of the TIM, as used in the numerical modelling, is shown in Fig. 7.16. Note that only two-dimensional modelling (x–y) was performed.

The experimental results (shown in brackets in Table 7.9) for comparison with the modeling results were obtained in prior work by sandwiching each coated sheet between two copper blocks at controlled pressure. The surface of each copper block has controlled roughness, such that two levels of roughness are evaluated.

In her earlier MS thesis (2003), Abadi had numerically modelled and compared the performance of two thixotropic thermal pastes, a carbon black paste (polyol esters vehicle with 2.4 vol% carbon black, and a metal particle paste (commercial Shin-Etsu X-23-7762 paste, which is aluminum particle filled silicone with density 2.6 g/ml, manufactured by Shin-Etsu MicroSi, Inc., Phoenix, AZ). Thixotropy refers to the ability of a paste to maintain its shape in the absence of an applied stress. In other words, the paste flows only under an applied stress. A thixotropic paste is attractive in avoiding seepage of the paste, say during transportation of the computer. Seepage may cause contamination in the electronic package.

Typical results are shown in Table 7.10. It can be seen that, although the metal particle paste is much more conductive thermally than the carbon black paste, the carbon black paste gave a better performance because it could be applied thinner (bond line thickness is smaller) than the metal particle paste, (see Teetstra (2007) discussed above) (Table 7.10).

The work of Leong et al. (2005) and Leong (2007) focused on the development of materials containing carbon as the thermally conductive component due to the ease of availability of carbon nanoparticles in the form of carbon black.

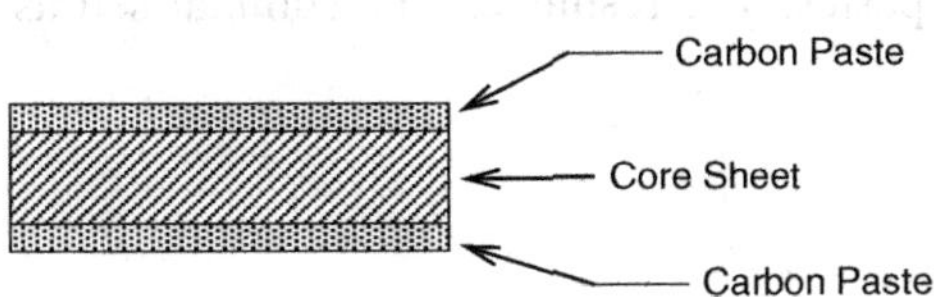

Fig. 7.16 Schematic of TIM used in numerical modelling by Abadi and Chung (2007)

Table 7.9 Modelling results for TCC for various combinations of core sheet and proximate surface roughness; experimental results are shown in parantheses (data from Abadi Chung (2007), used with permission from Springer)

		Rough proximate surface (15 μm)			Smooth proximate surface (0.01 μm)		
Pressure (MPa)		0.46	0.69	0.92	0.46	0.69	0.92
TCC (kW/m^2 K)	Aluminium	27.6 (35.4)	32.5 (38.5)	34.1 (41.9)	69.5 (62.2)	72.6 (69.4)	74.3 (70.9)
	Graphite	26.6 (23.4)	29.8 (26.0)	33.4 (29.4)	26.3 (30.9)	26.7 (32.1)	26.9 (34.8)
	Copper		21.5			60.8	
	Indium		36.8			78.7	

Note For aluminium and indium, the core sheet thickness was 7 μm; for copper, 13 μm; for graphite, 130 μm

Table 7.10 Thermal contact conductance (TCC), bond line thickness and fractional valley filling for various combinations of thermal paste type and copper surface roughness

Thermal paste	Roughness (μm)	Pressure (MPa)	TCC (10^4 W/m ^{2} K)*	TCC (10^4 W/m ^{2} K)†	Bond line thickness (μm)	Fractional Valley Filling (%)†
Metal particle	15	0.46	7.76	6.31	4.0	29.9
Metal particle	15	0.69	8.43	7.27	3.3	37.0
Metal particle	15	0.92	8.78	7.45	2.9	43.0
Carbon black	15	0.46	8.72	6.67	0.4	6.1
Carbon black	15	0.69	10.18	7.87	0.4	6.1
Carbon black	15	0.92	11.12	9.07	0.4	6.1
Metal particle	0.01	0.46	19.87	21.87	3.4	100
Metal particle	0.01	0.69	22.55	22.35	2.9	100
Carbon black	0.01	0.46	25.91	29.31	0.24	100
Carbon black	0.01	0.69	27.75	32.26	0.20	100

* Experimental results (from earlier work at SUNY, Buffalo)
† Modeling results

Nanoparticles are particularly effective for filling microscopic valleys in the surface topography, thereby enhancing the conformability of the paste. Nanofibers or nanotubes are not effective unless they are aligned in the direction of heat transfer, so their usage requires making them into an array of parallel nanofibers or nanotubes on a substrate. The array configuration limits the practical use of nanofibers or nanotubes, which are also much more expensive than carbon black.

The most important criteria for thermal interface material effectiveness are conformability, spreadability and thermal conductivity. The conformability, as noted earlier, is necessary to displace air from the interface. Spreadability is necessary to make the interface material thin. This work used PEG (polyethylene glycol) as the vehicle for fluid pastes and polyol ester as the vehicle for thixotropic pastes. The results can be summarised as follows:

- Carbon black (30 nm) thixotropic paste based on polyol esters is comparable to carbon black fluidic paste based on polyethylene glycol (PEG) in its effectiveness as a thermal paste, and in its dependence on pressure history.
- The optimum carbon black content is 2.4 vol% for the thixotropic paste.
- The thermal contact conductance across copper surfaces is 300×10^3 and 110×10^3 W/(m^{2}K) for surface roughness of 0.05 and 15 μm, respectively.
- Although possessing better conductivities, boron nitride (5–11 μm) and graphite (5 μm) thixotropic pastes are less effective than carbon black thixotropic paste due to their low conformability.

Fullem (2008) considered that TCR, specified on a unit area is more useful for characterising TIM bondlines since it takes area into account making it possible to compare bondlines of different areas.

He measured the thermal resistance of TIM bondlines composed of a commercially available electrically insulating, filled one-part epoxy (Eccobond E3503-1), Bondlines of several thicknesses were constructed by squeezing this material between two glass slides and then curing it. The results were compared with the performance of metal foils and are summarised (Table 7.11).

In his thesis, Hu (2010) explored the application of flexible graphite modified by carbon black paste as a TIM. Flexible graphite, which is made by compressing exfoliated graphite, is an attractive thermal interface material due to its resilience and thermal conductivity. Polyol-ester based carbon black (Cabot/Tokai) pastes may be used to coat or penetrate flexible graphite to increase its TCC.

Flexible graphite with carbon black thermal paste was prepared by compression of a column of exfoliated graphite in the presence of the thermal paste at the top and bottom surfaces of the exfoliated graphite column. The paste positioned directly above and below the exfoliated graphite column is in the form of a coating on a Teflon sheet. The two coated Teflon sheets (50 μm thick) are oriented perpendicular to the axis of the column. The compression is conducted in a cylindrical mold of length 45 cm and inner diameter 31.7 mm by applying a uniaxial pressure (11.2 MPa) via a matching piston. The pressure is held for 10 min. A part of the paste penetrates the resulting flexible graphite. The penetrant coats the surfaces of the "worms", rather than filling all the void spaces inside the flexible graphite sheet. The entire thickness of a flexible-graphite-based specimen is obtained in one

Table 7.11 Comparison of foils and epoxies (data from Fullem (2008))

Material	Thickness (μm)	Thermal contact resistance (10^{-4} m^2 K/W)	
Copper foil	26	335 at 415 kPa	230 at 800 kPa
Aluminium foil	14	175 at 225 kPa	105 at 740 kPa
Eccobond epoxy	226	330 at 650 kPa	
Eccobond epoxy	111	155 at 700 kPa	
Silver filled epoxy	29	10 at 465 kPa	

Note that Eccobond E3503-1 epoxy was tested with glass substrates whereas the rest were tested with silicon substrates. The effect of temperature does not seem to have investigated. The results confirm that BLT needs to be as small as possible for maximum benefit

11.2 MPa compression stroke. Each resulting specimen is a disc of diameter 31.7 mm.

In the case of paste-coated flexible graphite, one layer of thermal paste is applied by brushing (like painting) directly on each side of the plain flexible graphite specimen made beforehand at a pressure of 11.2 MPa without the presence of thermal paste in the column. The thickness of the coating is controlled by controlling manually the brushing force,

The minimum thickness of the flexible graphite is limited by the fragility of the specimens, which may fracture during the separation from the Teflon sheets after compression. The minimum thickness of plain flexible graphite is about 30 μm. With a coating of 10 μm thickness on each side, the *minimum* thickness of paste-coated flexible graphite is about 50 μm. When the 15 vol% Tokai paste is used as a penetrant, the minimum thickness is 42 μm. When the 2.4 vol% Cabot paste is used as a penetrant, the minimum thickness is 26 μm (Table 7.12).

It is seen that penetrating yields significantly higher values of conductance than mere coating. The increase of bond line thickness invariably results in reduced conductance.

Flexible-graphite-based thermal interface materials, being solid and therefore easy to handle, have an advantage over thermal pastes. Their lower tendency for seepage and larger thickness are their other advantages. The low tendency for seepage is due to the ability of the flexible graphite to retain the paste. A large thickness cannot be provided by a paste without a high tendency for seepage, but it is needed in practice for gap filling. The minimum thickness of flexible-graphite-based materials (30 μm) tends to be larger than that of metal-foil-based materials 7 μm. This gives the metal-foil-based material an advantage. The minimum thickness of silicone-based pads is 200 μm, which is much larger than that of flexible-graphite-based materials.The large thickness causes the silicone-based pads to exhibit very low values of the thermal contact conductance.

Aoyagi (2006) noted that the organic phase change materials, such as paraffin wax, are attractive in their low reactivity, stability in the phase change characteristics under thermal cycling but they tend to be poor in thermal conductivity.

As seen earlier, the thermal conductivity of a PCM can be increased by using a filler (particles, fibres or bars) that is thermally conductive. The filler does not melt, but its presence can affect the phase change characteristics, including the melting temperature and the heat of fusion. Due to the low thermal conductivity of the organic PCMs, the use of a thermally conductive filler is important. Aoyagi found that the conductance attained by BN filled diols to be higher than that attained by commercial phase change thermal interface materials.

The research of Howe (2006) was concerned with TIMs that were electrically resistive and thermally conductive, suitable for electronics cooling.

Materials in the graphite category are conductive both thermally and electrically. In contrast, diamond and boron nitride are thermally conductive and electrically resistive, thus making them attractive for thermal pastes that are electrically resistive—necessary to avoid short circuitng. However, diamond and boron nitride are expensive and are not available in the form of nanoparticles.

Table 7.12 Thermal contact conductance of selected thermal-paste modified flexible graphite specimens (based on the data of Hu (2010))

Composition	Thermal contact conductance, kW/(m^{2}K)		
	Thickness		
	50 μm	130 μm	300 μm
Plain	15.9	13.1	9.5
2.4 vol% Cabot coated	47.8	29	15.7
15 vol% Cabot coated	46.7	28.6	15.6
2.4 vol% Cabot penetrated	69	28.2	12.4
15 vol% Cabot penetrated	71.7	29.3	12.9

Electrically nonconductive thermal pastes were attained using carbon (5 μm graphite or 30 nm carbon black) as the solid component, together with electrically resistive particles (fumed alumina or exfoliated clay). The clay prior to exfoliation-adsorption processing is nanoclay (Cloisite 25A, from Southern Clay Products, Inc., Gonzales, TX). It is made of layered magnesium aluminum silicate platelets of thickness 1 nm and size 70–150 nm.

The relative performance was assessed by measuring the thermal contact conductance between copper surfaces (15 μm roughness) using the Guarded Hot Plate method. The following conclusions were drawn. Howe found that bBoth clay and alumina were highly effective in increasing the electrical resistivity for graphite pastes. However, clay was less effective and alumina was ineffective for carbon black pastes. With 1 vol% alumina and 10.7 vol% graphite, electrical resistivity of 10^{13} Ω.cm and thermal contact conductance of 88.3 × 10^3 W/m^2 °C were obtained. With 0.6 vol% clay and 10.7 vol% graphite, an electrical resistivity of 10^{12} Ω.cm and thermal contact conductance of 95 × 10^3 W/m^2 °C were obtained.

7.8.1 Summary: Other Interstitial Materials

In general, the bond line thickness of TIMs needs to be as small as practicable to obtain maximum enhancement in conductance. This implies that the adhesive pastes must be applied as thinly as possible. The desirable qualities of TIMs are:

- Conformability in order that the TIM is able to displace air from the interface gaps
- Spreadability is necessary to make the interface material thin
- Thermal conductivity

Nanoparticles, such as carbon black, are particularly effective for filling microscopic valleys in the surface topography, thereby enhancing the conformability of the paste. Nanofibres or nanotubes are not effective unless they are aligned in the direction of heat transfer, so their usage requires making them into an array of parallel nanofibres or nanotubes on a substrate. The array configuration limits the practical use of nanofibres or nanotubes, which are also much more expensive than carbon black.

Flexible-graphite-based thermal interface materials, being solid and therefore easy to handle, have an advantage over thermal pastes. An additional advantage is their lower tendency for seepage due to the ability of the flexible graphite to retain the paste. Penetrating the flexible graphite by the carbon black under pressure results in a more effective TIM than by merely coating it with the paste.

References

Abadi PP, Chung DDL (2011) Numerical modeling of the performance of thermal interface materials in the form of paste-coated sheets. J Electron Mater 40:1490–1500

Al-Astrabadi FR, O'Callaghan PW, Probert SD, Jones AM (1977) Thermal resistances resulting from commonly used inserts between stainless steel static bearing surfaces. Wear 40(3):339–350

Antonetti VW, Yovanovich MM (1985) Enhancement of thermal contact conductance by metallic coatings: theory and experiment. Trans ASME J Heat Transfer 107:513–517

Aoyagi Y (2006) Improving thermal interface materials through the liquid component. MS thesis, University of Buffalo, State University of New York

Biercuk MJ, Llaguno MC, Radosavljevic M, Hyun JK, Johnson AT (2002) Carbon nanotube composites for thermal management. Appl Phys Lett 80:2767 (12 pages)

Chung KC, Sheffield JW, Sauer HJ Jr, O'Keefe TJ (1993a) Thermal contact conductance of a phase mixed coating by transitional buffering interface. J Thermophys Heat Transfer 7(2):326–333

Chung KC, Benson HK, Sheffield JW (1993b) Thermal contact conductance of coated junctions within microelectronic packages. In: AIAA Paper 93-2774, AIAA 28th thermophys conference American Institute of Aeronautics and Astronautics. Washington, DC

Cividino S, Yovanovich MM (1975) A model for predicting the joint conductance of a woven wire screen contacting two solids. Prog Astronaut Aeronaut 39:111–128

Cola BA, Xu J, Fisher TS (2009) Contact mechanics and thermal conductance of carbon nanotube array interfaces. Int J Heat Mass Transfer 52:3490–3503

Cook RS, Token KH, Calkins RL (1982) A novel concept for reducing thermal contact resistance. In: AIAA/ASME joint conference, St. Louis. American Institute of Aeronautics and Astronautics, New York

Cunnington GR Jr (1964) Thermal conductance of filled aluminium and magnesium joints in a vacuum environment. In: ASME Paper 64-WA/HT-40. American Society of Mechanical Engineers, New York

Desai AH (2006) Thermal management of small scale electronic systems. PhD thesis, Binghamton University, State University of New York

Dryden JR (1983) The effect of a surface coating on the constriction resistance of a spot on an infinite half space. Trans ASME J Heat Transf 105:408–410

Fletcher LS, Miller RG (1973) Thermal conductance of gasket materials for spacecraft joints. Prog Astronaut Aeronaut 35:335–349

Fletcher LS, Smuda PA, Gyorog DA (1969) Thermal contact resistance of selected low-conductance interstitial materials. AIAA J 7(7):1302–1309

Fletcher LS, Cerza MR, Boysen RL (1976) Thermal conductance and thermal conductivity of selected polyethylene materials. Prog Astronaut Aeronaut 49:371–480

Fried E, Costello FA (1962) Interface thermal contact resistance problem in space vehicles. ARSJ 32:237–243
Fullem TZ (2008) Characterization of the heat transfer properties of thermal interface materials. PhD thesis, Binghamton University, State University of New York
Gao Y, Marconnet AM, Panzer MAS, Leblanc S, Dogbe S, Ezzahri Y, Shakouri A, Goodson KE (2010) Nanostructured interfaces for thermoelectrics. J Electron Mater 39:1456–1462
Gwinn JP, Webb RL (2003) Performance and testing of thermal interface materials, in thermal challenges in next generation electronic systems. Millpress, Rotterdam, pp 201–210
Gyorog DA (1971) Investigation of thermal isolation materials for contacting surfaces. Prog Astronaut Aeronaut 24:310–336
Howe T (2006) Evaluation and improvement of thermal pastes for microelectronic cooling. MS thesis, University of Buffalo, State University of New York
Hu K (2010) Flexible graphite modified by carbon black paste for use as a thermal interface material. MS thesis, University of Buffalo, State University of New York
Kang TK, Peterson GP, Fletcher LS (1990) Effect of metallic coatings on thermal contact conductance of turned surfaces. Trans ASME J Heat Transfer 112:864–871
Kharitonov VV, Kokorev LS, Tyurin YuA (1974) Effect of thermal conductivity of surface layer on contact thermal resistance. Atomnaya Energiya 36(4):308–310
Koh B, John JE (1965) The effect of foils on thermal contact resistance. In: Paper 65-HT-44, ASME-AIChE heat transfer conference. American Society of Mechanical Engineers, New York
Kshirsagar B, Nagaraju J, Krishna Murthy MV (2003) Thermal contact conductance of silicon nitride-coated OFHC copper contacts. Exp Heat Transf 16:273–279
Lambert MA, Fletcher LS (1992) A review of thermal contact conductance of junctions with metallic coatings and films. In: AIAA Paper 92-0709, AIAA 30th aerospace meeting, Reno, NV
Lambert MA, Fletcher LS (1993) A correlation for the thermal contact conductance of metallic coated metals. In: AIIA 28th thermophys conference paper AIAA 93-2778. American Institute of Aeronautics and Astronautics, Washington, DC
Leong C-K (2007) Improving materials for thermal interface and electrical conduction by using carbon. PhD thesis, Binghamton University, State University of New York
Leong C-K, Aoyagi Y, Chung DDL (2005) Carbon-black thixotropic thermal pastes for improvingthermal contacts. J Electron Mater 34:1336–1341
Li YZ, Madhusudana CV, Leonardi E (2000) Enhancement of thermal contact conductance: effect of metallic coating. Int J Thermophys Heat Transf 14:1–8
Madhusudana CV (1994) Control of thermal contact conductance—some practical considerations. In: Balakrishnan AR, Srinivasa Murthy S (eds) Heat and mass transfer. Tata McGraw-Hill, New Delhi, pp 183–188
Madhusudana CV, Villanueva EP (1996) Effect of interstitial materials on joint thermal conductance. In: Wilkes KE, Dinwiddie RB, Graves RS (eds) Thermal conductivity 23. Technomic Publishing, Lancaster, pp 553–563
Madhusudana CV, Man JKL, Fletcher LS (1996) Effective microhardness for the determination of contact conductance of coated surfaces. In: Proceedings of the 31st national heat transfer conference. ASME HTD 327:139–145
Marotta EE, Lambert MA, Fletcher LS (1994) Thermal enhancement coatings and films for microelectronic systems. In: Proceedings of the 10th international heat transfer conference, paper 15-CI-15. Institution of Chemical Engineers, Rugby, UK
Mikic B, Carnasciali G (1970) The effect of thermal conductivity of plating material on thermal contact resistance. Trans ASME J Heat Transf 92:475–482
Mohs WF, Madhusudana CV, Garimella SV (2000) Constriction resistance in coated joints. In: Proceedings of the 34th national heat transfer conference, Pittsburgh, PA, NHTC2000-12033, pp 1–7
O'Callaghan PW, Probert SD (1988) Reducing the thermal resistance of a pressed contact. Appl Energy 30:53–60

O'Callaghan PW, Snaith B, Probert SD (1983) Prediction of interfacial filler thickness for minimum thermal contact resistance. AIAA J 21(9):1325–1329

Ochterbeck JM, Fletcher LS, Peterson GP (1990) Evaluation of thermal enhancement films for electronic packages. In: Paper 17-Cd-05, 9th international heat transfer conference, Jerusalem. Hemisphere, New York, pp 445–450

Olsen E, Garimella SV, Madhusudana CV (2001) Modeling of constriction resistance at coated joints in a gas environment. In: 2nd International symposium on advances in computational heat transfer, Palm Cove, Qld, Australia, May 20–25

Olsen EL, Garimella SV, Madhusudana CV (2002) Modeling of constriction resistance in coated joints. J Thermophys Heat Transf 16:207–216

Peterson GP, Fletcher LS (1988) Thermal contact conductance in the presence of thin metal foils. In: AIAA Paper 88-0466. American Institute of Aeronautics and Astronautics, Washington, DC

Peterson GP, Fletcher LS (1990) Measurement of thermal contact conductance and thermal conductivity. Trans ASME J Heat Transf 112:579–585

Pop E, Mann D, Wang Q, Goodson K, Dai H (2006) Thermal conductance of an individual single-wall carbon nanotube above room temperature. Nano Lett 6:96–100

Prasher R, Chiu C-P, Mahajan R (2004) Thermal interface materials: a brief review of design characteristics and materials. Electron Cool 8:12 Pages

Rajamohan V, Madhusudana CV (1998) Characterisation of coated surfaces: prediction of the effective microhardness coated surfaces. In: Hargreaves D, Scott W (eds) Proceedings of the 5th international tribology conference, Queensland University of Technology, Brisbane, pp 519–523

Rajamohan V, Madhusudana CV (1999) Numerical model for microhardness testing. In: Hadi MNS, Schmidt LC (eds) Proceedings of the international conference on mechanics of structures, materials and systems, University of Wollongong, pp 207–213

Ramamurthi K, Sunil Kumar S, Abilash PM (2007) Thermal contact conductance of molybdenum-sulphide-coated joints at low temperature. J Thermophys Heat Transfer 21:811–813

Sauer HJ Jr (1992) Comparative enhancement of thermal contact conductance of various classes of interstitial materials. In: NSF/DITAC workshop, Melbourne, Monash University, Victoria, Australia, pp 103–115

Sauer HJ Jr, Remington CR, Stewart WE Jr, Lin JT (1971a) Thermal contact conductance with several interstitial materials. Proc Int Conf Therm Conduct 11:22–23

Sauer HJ Jr, Remington CR, Heizer G (1971b) Thermal contact conductance of lubricant films. Proc Int Conf Therm Conduct 11:24–25

Shaikh S, Lafdi K, Silverman E (2007) The effect of aCNT interface on the thermal resistance of contacting surfaces. Carbon 45:695–703

Sheffield JW, Chung KC (1992) Thermal contact conductance of metal to metal and metal to ceramic joints. In: NSF/DITAC workshop, Melbourne, Monash University, Victoria, Australia, pp 13–18

Sinha N, Yeow JTW (2005) Carbon nanotubes for biomedical applications. IEEE Trans Nanobiosci 4:180–195

Snaith B, O'Callaghan PW, Probert SD (1984) Interstitial materials for controlling thermal conductances across pressed metallic contacts. Appl Energy 16:175–191

Son Y, Pal S, Borca-Tasciuc T, Ajayan P, Siegel R (2008) Thermal resistance of the native interface between vertically aligned multiwalled carbon nanotube arrays and their SiO_2/Si substrate. J Appl Phys 103:024911 (7 pages)

Teertstra P (2007) Thermal conductivity and contact resistance measurements for adhesives. In: Proceedings of IPACK2007, ASME InterPACK '07. IPACK2007-33026, 8 pages

Villanueva EP (1997) Thermal contact conductance. PhD thesis, The University of New South Wales

Xu J (2006) Carbon nanotube array thermal interfaces. PhD Thesis, Purdue University

Xu, J, Fisher TS (2006) Enhancement of thermal interface materials with carbon nanotube arrays. Int J Heat Mass Trans 49:1658–1666

Yovanovich MM (1972) Effect of foils upon joint resistance: evidence of optimum thickness. Prog Astronaut Aeronaut 31:227–245

Yu C, Saha S, Zhou J, Shi L (2006) Thermal contact resistance and thermal conductivity of a carbon nanofiber. Trans ASME J Heat Trans 128:234–239

Chapter 8
Major Applications

In this chapter some major applications in which contact conductance plays a significant role will be discussed in detail. These applications include finned tube heat exchangers, a variety of manufacturing processes and heat transfer in stationary packed beds. The topics are chosen on the basis of their contemporary interest, practical significance and extensive information available in each category.

8.1 Contact Heat Transfer in Finned Tubes

Thermal contact resistance is a significant part of the overall resistance to heat transfer in finned tube heat exchangers, particularly those in which the joint between the tube and the fin is mechanical rather than metallurgical. In those exchangers where the mechanical coupling between the tube and the fin collar is achieved by expanding the tube, Sheffield et al. (1985) found that the thermal contact resistance could be more than 15 % of the overall resistance.

Among the finned tube heat exchangers whose performance is affected by thermal contact resistance, three primary types may be identified (Fig. 8.1). Often, the liner tube is made from corrosion resistant material such as stainless steel and the outer finned tube made from high conductivity materials such as aluminium or copper.

The works of Dart (1959) and Gardner and Carnavos (1960) represent early investigations of the heat transfer performance of interference fitted tubes. Gardner and Carnavos noted that, as the temperature increased, the fins expanded away from the tube wall. As the temperature further increased, a successively greater proportion of the heat was transferred through the gas gaps in the interface than through the actual metal-to-metal contact spots. Eventually, a point was reached where the gap between the fin base and the tube was opened to such an extent that the heat transferred through the solid spots was deemed to be zero and all of the heat was assumed to pass through the entrapped fluid. These observations were also supported by the calculations of Kulkarni and Young (1966). Piir et al. (2007)

C. V. Madhusudana, *Thermal Contact Conductance*,
Mechanical Engineering Series, DOI: 10.1007/978-3-319-01276-6_8,

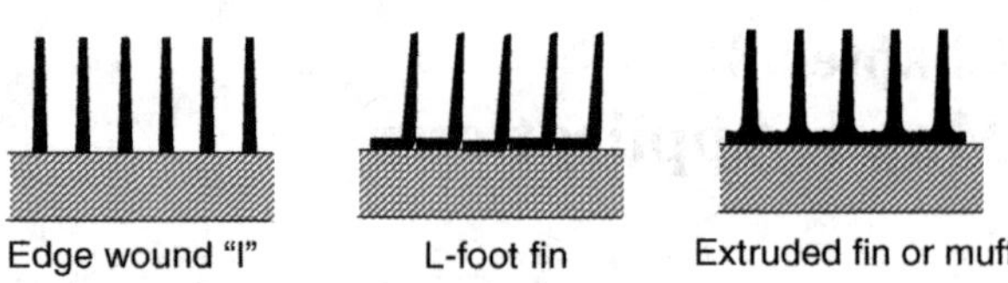

Fig. 8.1 Examples of fins in bimetallic finned tubes

recently investigated the effect of temperature cycling on bimetallic finned tubes (steel tube and aluminium shell). They observed that the TCR increased (TCC decreased) by a factor of 1.5 over the temperature range 50–260 °C. They further noted that although the steel tube was subjected to higher temperatures than the aluminum shell, it was compensated by the difference in the linear expansion coefficients of the materials. All of these results confirm the conclusions of the previous chapter (Sect. 7.2) on cylindrical joints.

The loss of contact sets a maximum limit for the operating temperature for specific types of heat exchangers. For example, Taborek (1987), while reviewing the status of bond resistance and design temperatures for finned tubes, noted that the following maximum bond temperatures were recommended (see Table 8.1).

We will consider briefly some of the recent investigations into the measurement of thermal contact resistance (and, sometimes, its variation) in finned-tube heat exchangers.

Deng et al. (1997) applied the method of parameter estimation to the measurement of the thermal contact resistance at the tube wall/fin collar interface. Tests were performed on an elementary heat exchanger consisting of two aluminium tubes joined by mechanical expansion to eleven aluminium fins. The associated thermal problem was also solved numerically by the finite element method. The thermal contact resistance was estimated by matching the computed fin temperature distribution with the measured temperatures and found to be 4.4×10^{-6} m^2 K/W and 2.7×10^{-5} m^2 K/W for the cold and the hot tube, respectively. The order of magnitude difference between the two resistances was attributed to the possible difference in the force used to expand the tubes into the fins.

Jeong et al. (2006) conducted an experimental investigation of fin-tube heat exchangers with various tube expansion ratios, spacing of fins, fin types (plate fin, slit fin and wide slit-fin) and manufacturing types of the tube (drawn tube and welded tube).

As shown in Fig. 8.2, the experimental apparatus consisted of vacuum chamber, vacuum pump, a pair each of constant temperature reservoirs, water pumps, mass flow meters, and thermo sensors. The fin-tube heat exchanger was composed of aluminium fins and grooved copper tubes with diameter of 7 mm, and had 12 tubes

Table 8.1 Maximum recommended bond temperatures. Maximum recommended bond temperatures

	USA	Europe
L-footed tubes	176 °C (350 °F)	150 °C (300 °F)
Extruded fins	230 °C (450 °F)	250 °C (480 °F)

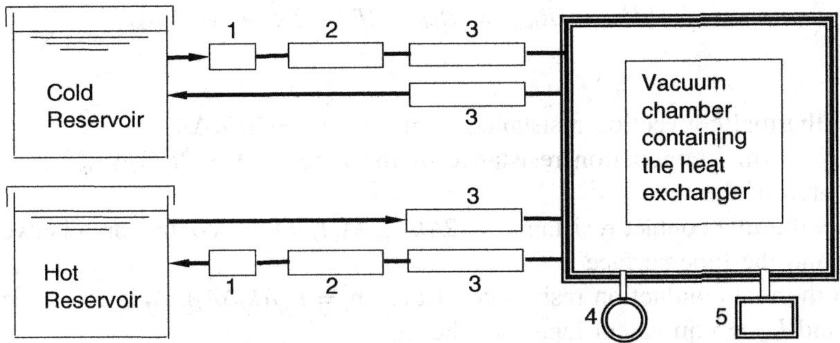

Fig. 8.2 Block diagram of the experimental apparatus used by Jeong et al. (2006): *1*—Pump; *2*—Flow meter; *3*—Temperature sensor; *4*—Vacuum gauge; *5*—Vacuum pump

Fig. 8.3 A fin-tube heat exchanger with 7 mm tube used by Jeong et al. 2006 (Reproduced by permission from Elsevier)

in a row as depicted in Fig. 8.3. Inlet and outlet tubes of cold water were located side-by-side in the upper part of the chamber wall and those of hot water placed in the lower part, to minimize the heat transfer from hot to cold water through the chamber wall. Also, the fin-tube heat exchanger was placed in an insulated vacuum chamber ensuring that the aluminium fins function only as a conduction medium with a minimum of natural convection on their surface.

The resistances to be considered are shown in Fig. 8.4. Considering an area element, ΔA, the heat flow rate is given by

$$dQ = U(T_h - T_c)\Delta A \tag{8.1}$$

Fig. 8.4 Resistances associated with the heat flow in the finned tube heat exchanger

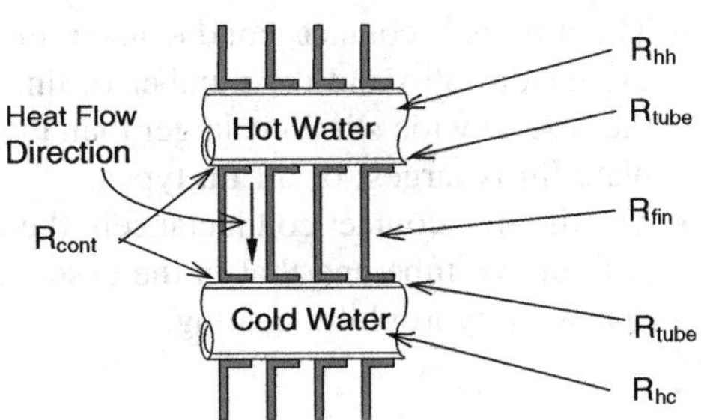

$$1/(UdA) = R_{\text{total}} = R_{hh} + 2R_t + 2R_c + R_f + R_{hc} \tag{8.2}$$

where

R_{hh} = thermal convection resistance of hot water = $2/(h_h \Delta A)$
R_{tube} = thermal conduction resistance of the tube wall = $2t\ (k_t \Delta A_m)$; A_m is log mean area of the tube
R_{cont} = thermal contact resistance = $2/(h_{\text{cont}} dA_c)$; dA_c = contact area between fin collar and the tube surface
R_{fin} = thermal conduction resistance of the fin = $l_{eq}/(k_f \Delta A_f)$; A_f is cross sectional area, and l_{eq} = equivalent length of the fin
R_{hc} = thermal convection resistance of cold water = $2/(h_c \Delta A)$.

A numerical calculation was used to evaluate quantitatively a thermal contact conductance which allows both the measured and the numerical heat balances to be the same. This involved carrying out, the numerical calculation repeatedly, changing the presumed value of thermal contact conductance, until the computed outlet temperatures of the hot and cold water equalled those of the experiment within a prescribed error.

The correlation for thermal contact conductance derived by Jeong et al. is given below (*Note* this correlation applies to tested configuration of 7 mm dia tubes only):

$$h\left[\frac{W}{m^2K}\right] = 893.7\left(\frac{t_f^3}{(P_t - t_f)^2 D_o}.E.S_f.M_t.C_f \times 10^4\right) + 899.0 \tag{8.3}$$

where

t_f = thickness of fin (m)
P_t = spacing (pitch) of the fin (m)
D_o = outer diameter of the tube
E = expansion ratio = $(D_{\text{ball}}/D_{\text{min}} - 1) \times 100$; (5.19, 6.01 and 6.82, corresponding to D_{ball} = 6.48, 6.53 and 6.88 respectively, in the tests)
S_f = 1 for slit fin; 1.25 for wide slit fin; 1.28 for plate fin
M_t = 1 for drawn tube; 1.12 for welded tube
C_f = 1 for without coating; 1.11 for with coating.

The authors arrived at the following conclusions:

- The thermal contact conductance increases with the increase of the tube expansion ratio and the number of fin, and the thermal contact conductance in the case of wide slit fin is larger than that of normal slit fin and that in the case of plate fin is largest of all fin types.
- The thermal contact conductancein the case with welded tube is larger than that with drawn tube and that in the case without hydrophilic coating is larger than that with hydrophilic coating.

- The portions of the thermal contact resistance are about 15–25 % in cases of the fin-tube heat exchanger with 7 mm tube, and it implies that the thermal contact resistance may not be ignored in the process of design of the fin-tube heat exchanger.

Note It is interesting to note that there is a finite thermal contact conductance term associated with welded tubes also. In other words, contact resistance is not zero for welded joints as normally assumed. Indeed Zhao et al. (2009) recently investigated compact heat exchangers in which copper fins were brazed to brass tubes. A microscopic study demonstrated that the actual brazed joints (even for a state-of-the-art manufacturing process performed under tightly controlled conditions) do have significantly poorer thermal integrity than an ideal joint. This indicated that the assumption of zero contact resistance is not valid, as assumed in a conventional thermal design procedure. They also noted that the impact of contact resistance on the heat transfer process may be even more significant for stainless steel (SS) and Ti exchangers due to stringent brazing conditions and significantly smaller thermal conductivity of the joint fillet.

According to the conclusions and Fig. 8.5 (Fig. 9 in Jeong et al. 2006), the contact conductance without coating should be higher than the one with coating. However, the factor C_f in the correlation seems to imply that it should be higher with coating!

In the majority of previous work on finned tube heat exchangers, the TCC is assumed to be a constant for a given exchanger. From the previous discussion on cylindrical joints, however, it is clear that in fin-tube heat exchangers the conductance should also depend on the heat flow rate and, in general, is not a constant.

Also, in all of the investigations reported so far, the contact conductance has been measured *indirectly*, that is, it is estimated as the difference between the total resistance and the sum of all other resistances. The other resistances, themselves, were either measured or calculated. Cheng and Madhusudana (2004, 2006a, b) carried out several experimental investigations in which the contact conductance was measured directly. This involved measuring the temperatures at several locations on the tube and the fin.

Photographs of the apparatus and the test section of the heat exchanger may be seen in Fig. 8.6. Each test specimen was assembled by mechanically expanding

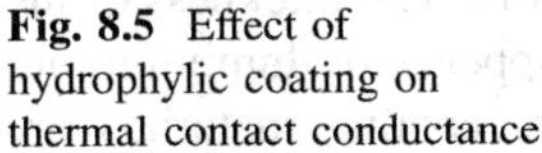

Fig. 8.5 Effect of hydrophylic coating on thermal contact conductance

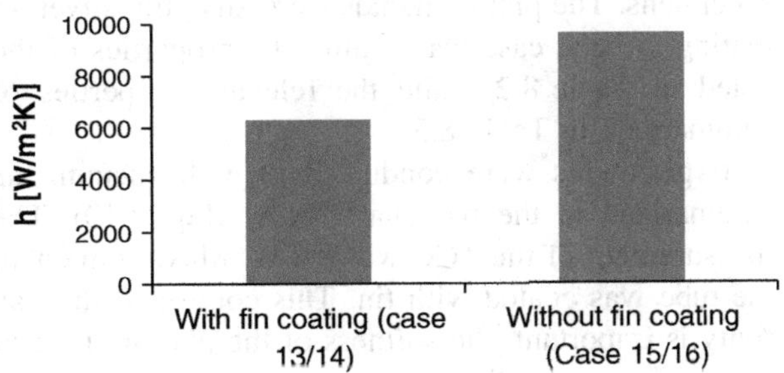

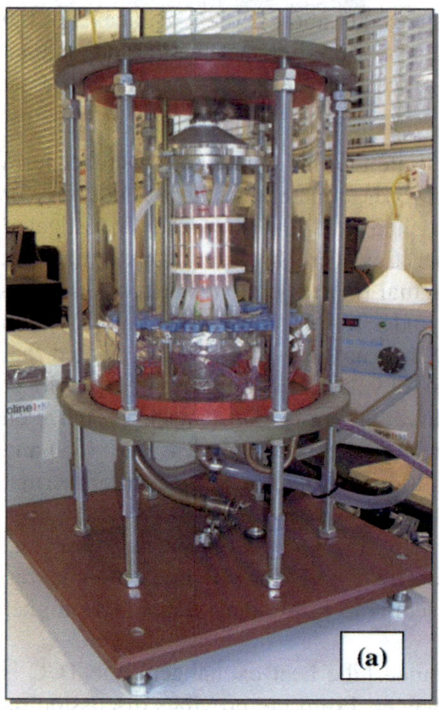

Fig. 8.6 **a** View of the experimental apparatus with the test section in place. **b** The test section with seven *tubes* and a fin sheet (Cheng 2006)

Fig. 8.7 Mechanical expansion *bullet* and *copper tube* (Cheng 2006)

seven copper tubes into a single aluminium fin. The expansion bullet and the copper tube are shown in Fig. 8.7. Shown in Fig. 8.8 are the locations of the thermocouples. The schematic diagram of the entire apparatus is as shown in Fig. 8.9.

The specimens included one non-coated specimen and four electroplated specimens. The plating metals were zinc, tin, silver and gold. The thickness of the plating in each case was 5 μm. The properties of the copper and aluminium are listed in Table 8.2 while the relevant properties of the coating materials are summarized in Table 8.3.

Experiments were conducted in both vacuum and nitrogen. The results are summarized in the bar chart below (Fig. 8.10). The overall uncertainty in the measurement of the TCC was 8.3 %. Maximum enhancement was obtained when the tube was coated with tin. This confirmed that, although the thermal conductivity is important, the softness of the plating material plays an important part in

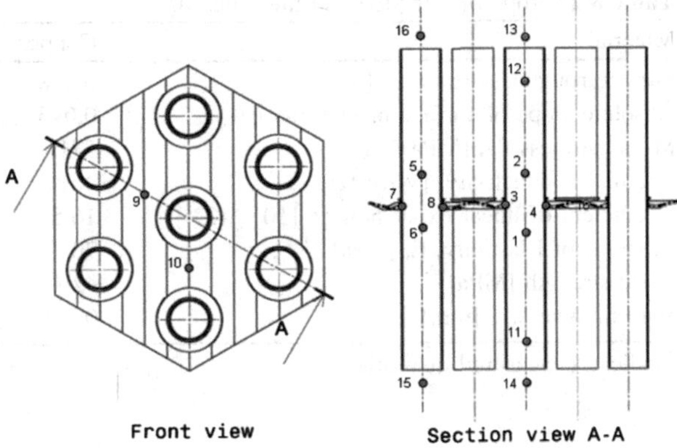

Fig. 8.8 Location of thermocouples on the fin and the *tubes* (Cheng 2006)

enhancing the TCC. The presence of an interstitial gas, such as nitrogen, is beneficial for the heat transfer and the TCC.

In another series of experiments, Madhusudana and Cheng (2007) investigated the effect of using bullets of different size in expanding the copper tubes and also the effect of heat flux on the contact pressure and contact conductance. Each specimen was tested in an atmosphere of nitrogen. Nitrogen was used to represent "normal" atmospheric conditions while avoiding the effect of variable moisture present inatmospheric air and also to prevent any oxidation.

The solid spot conductance was determined by subtracting the gap conductance from the measured total conductance. It was observed that while the range of the total conductance, for all specimens, was from 228 to 619 kW/(m^2 K), the range for the gap conductance was from 23.2 to 25.9 kW/(m^2 K). In other words, the

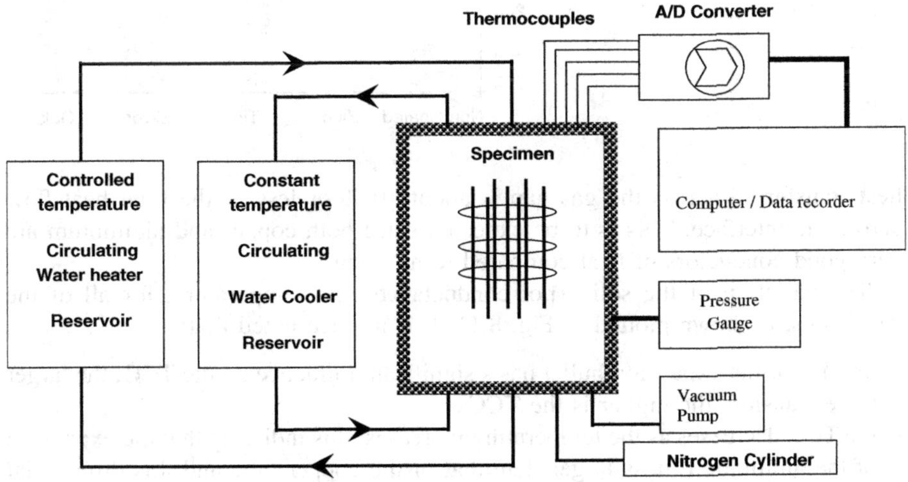

Fig. 8.9 Schematic diagram of the experimental setup (Cheng 2006)

Table 8.2 Properties of tube and fin materials

Material	Copper (bare tube)	Aluminium (fin)
Surface roughness (rms) σ, [μm]*	0.338	0.275
Absolute slope of surface profile, tan θ*	0.046	0.021
Microhardness, H, [MPa]*	571	287
Thermal conductivity, [W/(mK)]	401	237
Coefficient of linear expansion, α, [10^{-6} K^{-1}]	16.5	23.1
Modulus of Elasticity, E, [GPa]	70	110
Yield strength [MPa]	420	250
Wall thickness, t, [mm]	0.31	0.12

*Indicates measured quantities

Table 8.3 Properties of coating materials

Plating material	Thermal conductivity (Wm^{-1} K^{-1})	Microhardness, MPa
Zinc	116	675
Tin	67	199
Silver	429	626
Gold	317	1194

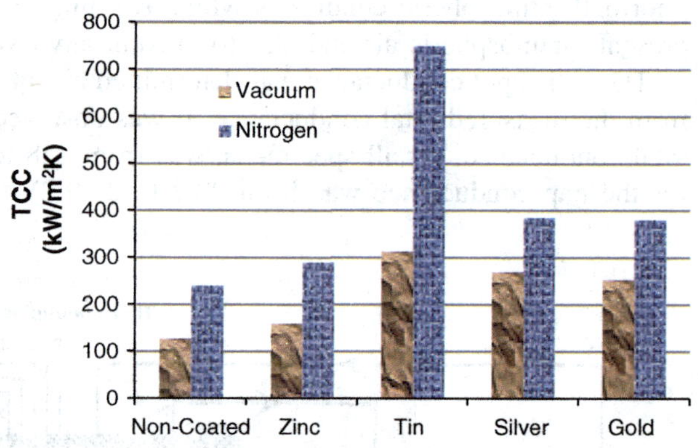

Fig. 8.10 Experimental results for thermal contact conductance of finned *tubes*

heat transferred across the gas gap is about 10 % or less of the total heat flow across the interface. This is to be expected since both copper and aluminium are very good conductors of heat compared to nitrogen.

The variation of the solid spot conductance with temperature, for all of the specimens, is shown plotted in Fig. 8.11. It is at once noted that:

- the size of the expansion bullet has a significant influence on the TCC; the larger the expansion, the higher is the TCC
- the TCC decreases as the temperature increases; this indicates that the expansion of the aluminium fin is larger than that of the copper tube and this differential

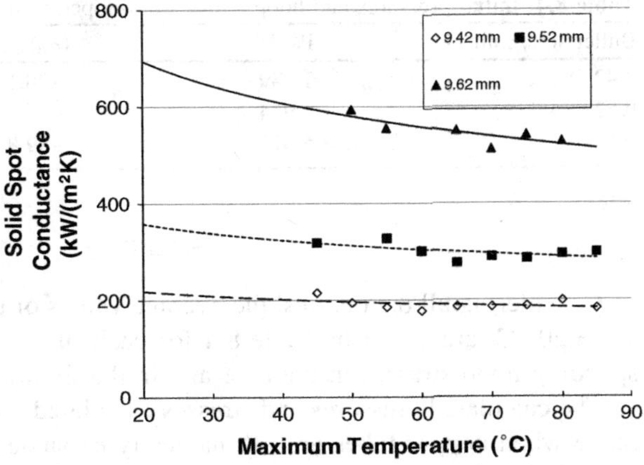

Fig. 8.11 Finned tube heat exchanger: variation of solid spot conductance with maximum temperature; effect of varying the size of expansion *bullet* is also shown

expansion causes the interface pressure (obtained at the time of assembly by the mechanical expansion of the tube into the fin) to relax gradually.

From the thermal conductance data as presented above, the corresponding variation in contact pressure may be computed using Tien's correlation, Eq. (3.43)
.

$$h_s = 0.55 \tan\theta\left(\frac{k}{\sigma}\right)\left(\frac{P}{H}\right)^{0.85}$$

The estimated contact pressures are shown plotted against temperature in Fig. 8.12. As the temperature was increased from 45 to 85 °C, the pressure decreased from 1.74 to 1.44 MPa, from 2.93 to 2.45 MPa, and from 5.58 to 5.05 MPa for the 9.42, 9.52 and 9.62 mm bullets respectively.

The shrink-fit pressure, p_0 at the time of assembly may be estimated by extrapolating the pressure–temperature graphs to 20 °C (assembly temperature). Then the initial hoop stresses, σ_θ in the tube and the collar may be calculated from the formula for thin tubes:

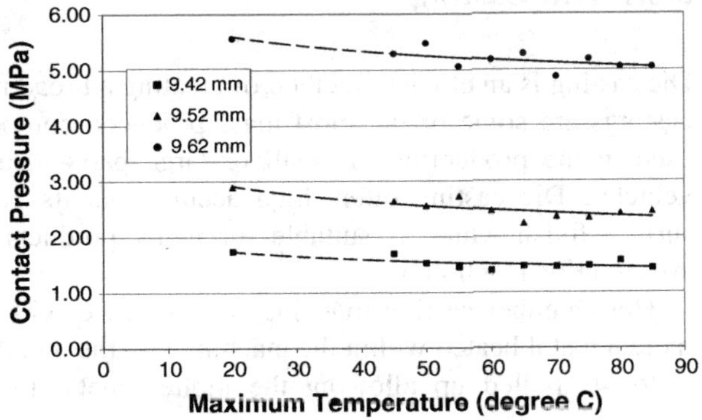

Fig. 8.12 Finned tube heat exchanger: Variation of contact pressure with maximum pressure, also showing estimated initial pressure for each bullet size

Table 8.4 Initial pressures and hoop stresses developed in the three tube-fin specimens

Bullet size, mm	P_0, MPa	$(\sigma_\theta)_{tube}$, MPa	$(\sigma_\theta)_{fin}$, MPa
9.42	1.739	−28.1	72.5
9.52	2.944	−46.8	120.8
9.62	5.615	−89.8	232.1

$$\sigma_\theta = Pr/t$$

Considering all data points, the average values of the initial contact pressure, at $T_0 = 20$ °C, are given in Table 8.4 for each of the three specimens. The corresponding hoop stresses in the tube and in the fin are also shown in this table.

The calculated pressures and stresses calculated also appear to be in the correct range when copper tubes are mechanically expanded into aluminium fins. However, for the 9.62 mm expansion, the hoop stress in the aluminium is high and is close to the yield strength of the material. This is likely to damage the fin material. In fact, there were some problems encountered in the manufacturing process using the 9.62 mm bullet. These included the difficulty in holding the tube during the expansion process and occurrence of fractures on the tube wall surface.

8.2 Manufacturing Processes

There are several manufacturing processes in which TCR plays an important role in the efficiency of the process and the ultimate quality of the product. Discussed briefly below are some representative processes such as die casting, injection moulding, resistance spot welding and thermal spray painting. Because of the very nature of the manufacturing processes, the temperature and heat flux measurements are necessarily associated with transient states. Any corresponding special or unique experimental techniques are also described.

8.2.1 Die Casting

Die casting is an efficient method of creating a broad range of complex shapes. Die castings are some of the most mass produced components today. Die casting is used in the production of small or large parts—from toy cars to parts of real vehicles. Die casting offers high accuracy in its products with a good quality surface finish which is suitable for many products without the need for extra polishing or machining.

Hot chamber casting machines use an oil or gas powered piston to drive the molten metal heated within the machine into the die. At the start of the process, the piston is pulled up allowing the molten metal to fill the "goose neck" (see

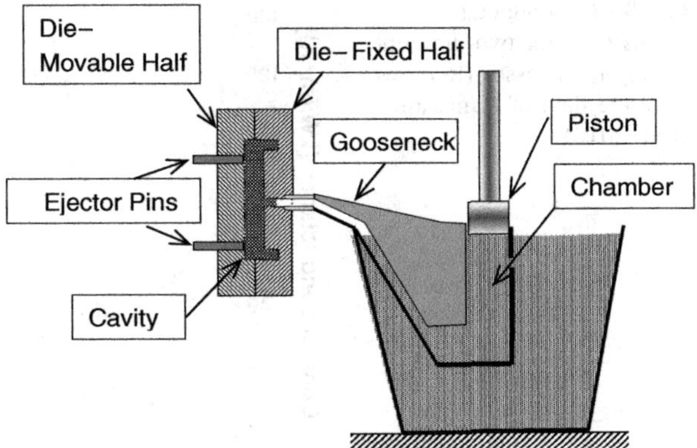

Fig. 8.13 The die casting arrangement

Fig. 8.13). Once the liquid metal has filled, the goose neck the piston is pushed down to force the liquid metal into the die. In High Pressure Die Casting (HPDC), the molten metal is injected at high velocity (50–60 ms^{-1}). Pressure of the order of 300–1000 bars is applied and maintained until molten metal is completely solidified in the die. The casting (product) is then ejected. This process has fast cycle times, e.g., a couple of seconds.

The solidification rate is governed by the magnitude of the thermal conductance at the casting–die interface. Experiments have shown that value of the conductance varies with solidification time.

Hamasaiid et al. (2010) reported the test results and analytical predictions for two different alloys, AZ91D (a die casting alloy with approximately 90 %Magnesium and 9 % Aluminium) and Al-9Si-3Cu, (a die casting alloy with approximately 85 %Aluminium and 9 % silicon, 3 % copper).

When a molten alloy is brought into contact with a relatively cold die and heat is transferred from the melt to the die, nucleation and grain growth should begin at the contact areas, and more precisely, around the peaks of the asperities. In the case of HPDC, the large applied pressure on the casting produces a large maximum value for the thermal conductance at the first stage of contact, when the molten alloy is still mostly liquid. When the density of contact spots and the conductance are large, the nucleated grains around the peak of the asperities in contact reach each other quite rapidly and form the first solidified film of the casting (casting skin) near the interface. Hence, the liquid–solid contact is transformed almost immediately into a situation of solid–solid contact. When the casting skin forms, the thermal gradient in the solidified film causes local shrinkage and possibly defects in the solidified component. This contraction may cause the solidified skin to partially separate from the die if the main contraction direction is perpendicular to the interface mean plane. Because of the rigidity of both solids and of the slope of the asperities, the interface necessarily moves apart, resulting in the reduction of the number and the area of the micro-contact spots. This is the main reason for the sharp decrease in conductance as a function of solidification time (see Fig. 8.14).

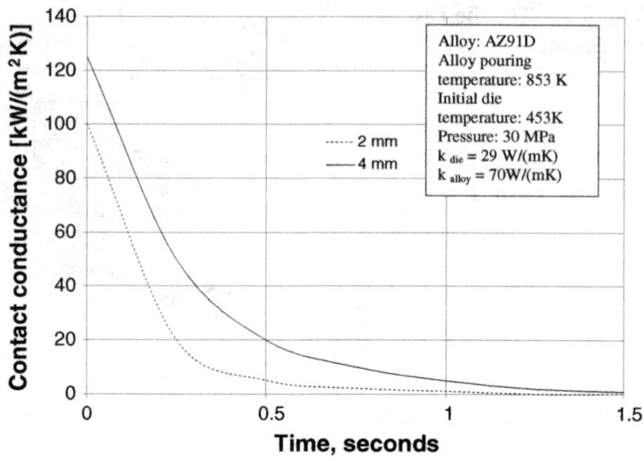

Fig. 8.14 Conductance versus time for two different casting thicknesses (Redrawn from the data of Hamasaiid et al. 2010)

8.2.2 Injection Moulding

Although injection moulding may be performed with most materials including metals and glasses, thermoplastic and thermosetting plastics are most commonly used for manufacturing products by this process.

In injection molding, the solid plastic material is first heated until it becomes a fluid (Fig. 8.15). This melt is then transferred under pressure (injected) into a closed hollow space (mold cavity) and then cooled in the mold cavity until it again reaches the solid state (below its glass transition or crystallization temperature). Following the cooling period, the mold is opened and the part is ejected from the cavity. A typical molding cycle thus consists of four stages: filling, packing, cooling and ejection.

1. Granulated or powdered thermoplastic plastic is fed from a hopper into the Injection Moulding machine.
2. The Injection Moulding machine consists of a hollow steel barrel, containing a rotating screw (Archemidial Screw). The screw carries the plastic along the barrel to the mould. Heaters surround the barrel melt the plastic as it travels along the barrel.
3. The screw is forced back as the melted plastic collects at the end of the barrel. Once enough plastic has collected, a hydraulic ram pushes the screw forward injecting the plastic through a sprue into a mould cavity. The mould is warmed

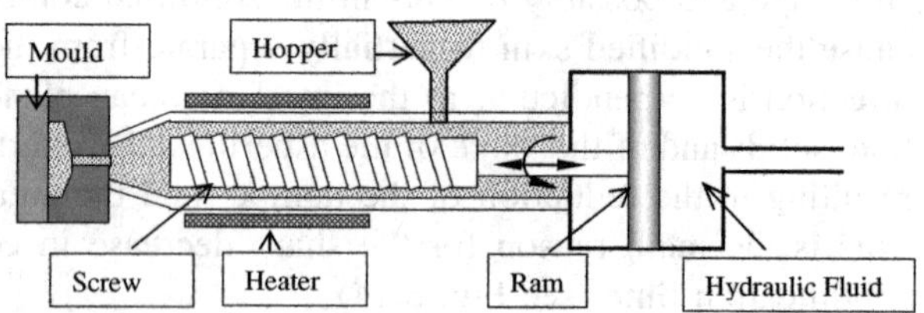

Fig. 8.15 The essentials of an injection moulding machine

before injecting and the plastic is injected quickly to prevent it from hardening before the mould is full.

4. Pressure is maintained for a short time (**dwell time**) to prevent the material creeping back during **setting** (hardening). This prevents shrinkage and hollows, therefore giving a better quality product. The moulding is left to cool before removing (**ejected**) from the mould. The **moulding** takes on the shape of the mould cavity.

The use of an accurate value of TCR between the plastic and the mould wall in the thermal simulation software for injection moulding will contribute to accurate results that can lead to economical design and a higher quality product. During a typical injection moulding process for thermoelastic polymers, the cooling stage comprises up to 75 % of the total cycle time. Li (2004) investigated the role of TCR in the injection mould cooling associated with the production of poly ethylene tetraphthalate (PET) products. In general, TCR increased with time during the cooling period due to shrinkage which caused the outer surface of the part to move away from the inner surface of the mould, and a gap to form. Higher holding pressures and injection rates delayed the gap formation, decreased the shrinkage and the TCR. Reduced cooling times and total cycle times are the benefits of a reduction in the contact resistance.

The experimental investigation of Bendada et al. (2003) into injection moulding was based on the combination of two non-invasive sensors to characterize thermal contact resistance TCR.

TCR was defined on a *unit area basis*:

$$TCR = (T_{ps} - T_{ms})/q,$$

where T_{ps} is the polymer surface temperature, T_{ms} is the mold surface temperature, and q is the heat flux crossing the interface.

To determine polymer surface temperature T_{ps}, Bendada et al. used a hollow waveguide pyrometer. Mold surface temperature T_{ms} and heat flux density q were indirectly obtained with the use of a specially designed two-thermocouple probe, Fig. 8.16. The latter was composed of two steel half cylinders joined side by side. These were obtained by cutting longitudinally a cylinder that was 8 mm in diameter and 13 cm long. The shape and size of the cylinder tip planned to be in contact with the polymer stream were designed to fit commonly employed probe-housing cavities in injection moulds. Two E-type fine-wire thermocouples 75 μm in diameter were spot-welded inside the probe at two different locations (1 and 2 mm) from the probe tip. At the interface between the two half-cylinders, a narrow slot was longitudinally machined in one half-cylinder to contain the thermocouples wires. The shape and the location of the slot are shown inset in Fig. 8.16. To perform measurements that were as nonintrusive as possible, the two-thermocouple probe was manufactured with the same steel (P20 steel grade) as the mold material and the same roughness as the cavity surface.

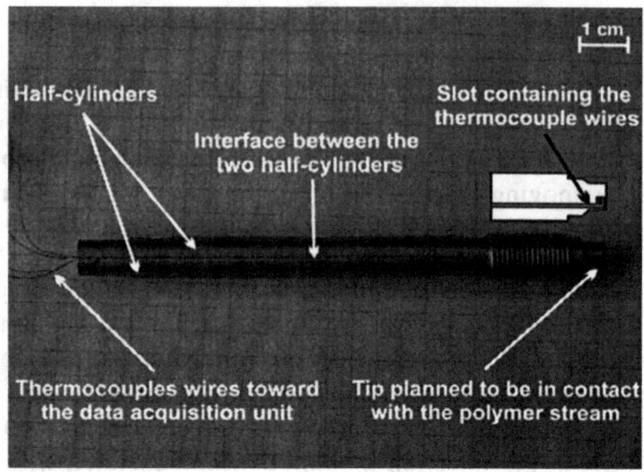

Fig. 8.16 Image of the two-thermocouple probe (Bendada et al. 2003, Reproduced with permission from American Institute of Physics)

From temperature histories monitored with the two-thermocouple probe, a regularized sequential inverse method allowed the estimation of both heat flux q crossing the polymer-mold interface and temperature at the cavity surface *T*ms.

Bendada et al., noted that at short times, TCR did not change as long as high pressure was maintained inside the cavity. When the cavity pressure dropped back to zero, a sudden rise in TCR was observed. The sudden rise of TCR was related to the appearance of the air gap caused by the polymer shrinkage.

Investigation of thermal contact resistance at the interface between the metal and the plastic in injection moulding was the topic of the PhD dissertation of Sridhar (1999). In his work, an inverse method was used to determine the thermal conductivity and TCR from transient measurements. Typical parameters used in the analysis were as follows:

- Injection pressure: 12 MPa
- Fill time: 1.1 s
- Melt temperature: 218 °C
- Cooling time (estimated): 52 s
- Mould temperature: 60 °C

The results were also discussed in a subsequent paper by Sridhar et al. (2000). It was found that as long as the cavity pressure was above atmospheric, the internal pressure that kept the part surface in good contact with the mould wall. The gap between the part and the mould wall was ssumed to start forming once the cavity pressure dropped to zero. Figure 8.17 shows the calculated thickness gap as well as the gap resistance generated as a result of shrinkage alone, during the post-filling period. The resistance reaches a maximum value of 0.0016 m^2 K/W at ejection. This is comparable to the value of 0.0009 m^2 K/W proposed by Yu et al. (1990) for a polystyrene part of 4 mm thickness.

Note Although the Eq. (4.4) for gap conductance

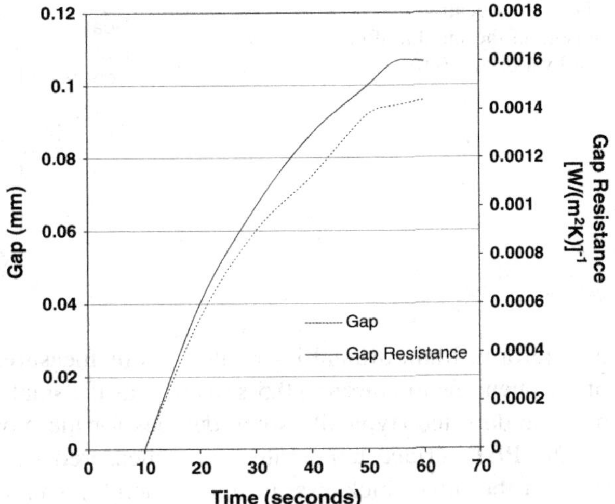

Fig. 8.17 Gap between plastic surface and mold wall and gap thermal resistance plotted as a function of post filling time (graphs re-plotted from data of Sridhar 1999)

$$h_g = \frac{k_g}{\delta + g_1 + g_2}$$

was quoted, no temperature jump distances appear to have been used in the analysis. However, the gap thicknesses are too large to be affected by temperature jump distances.

8.2.3 Stretch Blow Moulding

During a stretch-blow moulding cycle, the PET preform is successively blown and maintained into contact with an aluminium mould, under air pressure. The polymer is cooled down by the die, and it rigidifies inside the cavity. In such process, the solidification time of the part is controlled by the heat transfer between the plastic and the mould. Although the time of contact is very short (typically around 0.5 s), the heat transfer affects the mechanical properties of the bottle, and the quality of final parts. The PET cooling rate also affects the process efficiency.

For these reasons, the PET cooling rate prediction is of prime interest, and cannot be carried out without a precise understanding of the heat transfer properties. Owing to the imperfect nature of contact between plastic part and mould inner surfaces, the heat transfer is affected by TCR, some value of which is generally used in numerical software in order to impose the heat flux boundary condition. Although its value is critical to achieve accurate simulations, it is often poorly estimated.

With this in mind, the Bordival et al. (2007) developed special sensors in order to measure at the same time, the evolution of the TCR and the air pressure inside

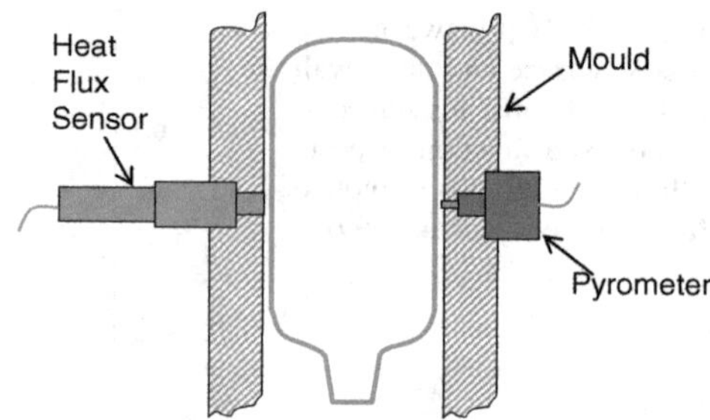

Fig. 8.18 Location of sensors in the mould, after Bordival et al. 2007

the preform. Main difficulties related to this measurement are the very short time of the phenomenon (around 0.5 s) as well as the small temperature increases which must be detected (typically some degrees for the mold).

The PET surface temperature was measured by infrared pyrometer, inserted through the mold thickness, in middle-height. This type of measurement has the advantage of being non-intrusive. The heat flux sensor was inserted in the same position as the pyrometer, but on other side of the mould (see Fig. 8.18).

The peak value of the heat transfer coefficient (averaged on five trials) was measured to be 225 W/(m^2 K), which corresponds to a TCR equal to 0.0044 m^2 K/W. The standard deviation of this peak value was 20 W/(m^2 K), (9 % error).

8.2.4 Rapid Contact Solidification

The interfacial heat transfer between the melt and the substrate is a critical issue for many industrial applications based on molten droplet impact, such as thermal spray coating, splat quenching, solder jet printing, high precision net-form manufacturing and shape deposition manufacturing (SDM) with micro-casting. During a melt spread over a solid surface, a perfect thermal contact between the solidifying metal and the solid substrate can hardly be achieved because of the roughness of the solid surface, the surface tension of the melt, the impurities on the surfaces and the gas entrapment. This imperfection leads to TCR that reduces the heat transfer rate.

Hong and Qiu (2007) performed a one-dimensional rapid contact solidification experiment by impacting a substrate onto a thin layer of molten metal to simulate the local contact heat transfer and solidification in molten droplet impact process. The effect of existence of the lateral flow on TCR during molten droplet impact was assumed to be negligible. This assumption was considered to be reasonable because in most of the molten droplet impact processes, the droplet impact velocity is small (≈ 1 m/s) and the viscosity of the molten metal is high. The variable TCR is estimated by solving an inverse heat transfer problem based on the

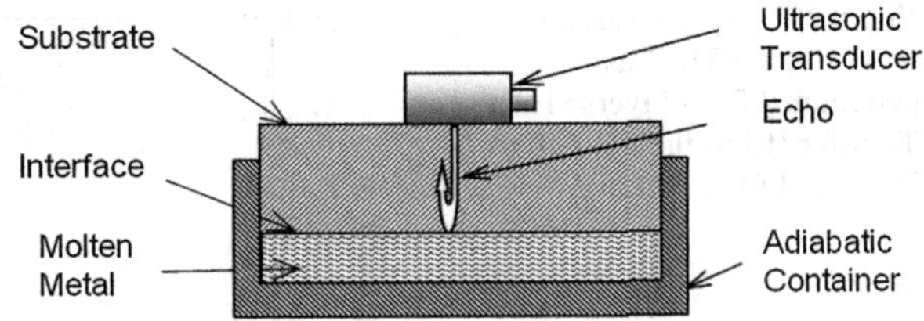

Fig. 8.19 Measurement technique used by Hong and Qui for the one-dimensional solidification problem

measurements of the delay time of ultrasound echoes from the contact interface, see Fig. 8.19.

Because the speed of sound in a material is a function of temperature, the delay time measured will carry the temperature information along the ultrasound path which can be used to determine the variation of thermal contact resistance.

The relationship between the speed of sound, c and the temperature of the medium, T can usually be expressed as

$$c(T) = c_0 + \gamma T$$

where c_0 is the speed of sound at 0 °C (m/s) and γ is the temperature coefficient (m/s/°C).

As, the substrate is heated continuously after contact with the molten metal, the measured time delay of ultrasonic echoes from the contact interface increases correspondingly. The change of the *measured* time delay of the echo (*UM*) after contact can be derived by subtracting the initial time delay of the last echo before contact from the time delay of the echo after contact.

On the other hand, supposing the variation of TCR with time is known, the temperature distribution inside the substrate can be computed by solving the heat transfer problem. The *computed time* delay change (*UC*) of the echo can then be calculated as

$$UC = \int_0^a \frac{dz}{c[T(z)]} - \frac{2a}{c(T_{s,0})}$$

where, $T(z)$ is the temperature along the sound path, $T_{s,0}$ is the initial substrate temperature, and a is the thickness of the substrate.

The variation of TCR is estimated by iteration, such that the difference between *UC* and *UM* is minimized. In other words, by comparing *UM* with *UC*, the variable TCR can be estimated using the inverse heat transfer method.

Compared with traditional inverse methods using the thermocouple to measure temperature, the ultrasonic method has some advantages. Firstly, it does not affect the heat transfer process inside the substrate, because no thermocouple is buried inside the substrate. Therefore, this method is suitable for thin layers of molten liquid and substrate impact processes. Secondly, the time delay of the ultrasonic echo actually reflects the integration of the temperature field along the sound path, revealing the temperature change at any location of the substrate. Therefore the

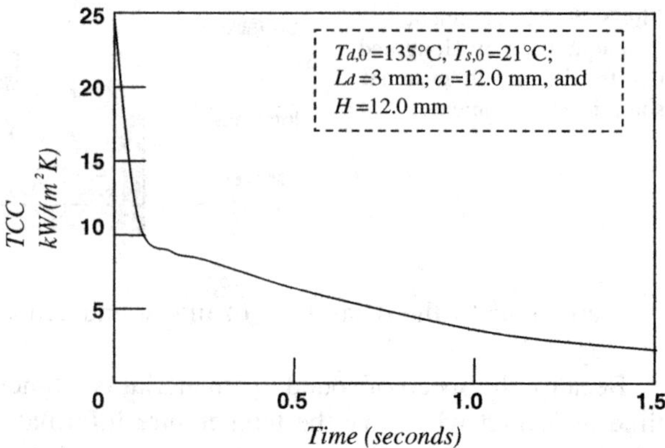

Fig. 8.20 Typical variation of TCC with Time as determined from Inverse Heat Transfer (from the data of Hong and Qui, 2007)

well-known problem of the lagged temperature response in the interior point of the substrate with respect to the heat flux excitation at the contact interface is avoided.

Hong and Qui used Alloy158, [an alloy of Bismuth (50 %), Lead (26.7 %), Tin (13.3 %) and Cadmium (10 %). the 158 refers to its melting point which is 158 °F (70 °C)] for both the substrate and the molten phase. Typical results are shown in Fig. 8.20 (for an impact height of 12 mm). It is seen that the interfacial heat transfer coefficient decreases rapidly at the first stage after the contact, the decrease becomes less noticeable thereafter.

Note In the above graph, $T_{d,0}$ = Initial Temperature of Molten Metal; $T_{s,0}$ = Initial Temperature of Substrate; L_d = Thickness of Molten Metal; a = Thickness of Substrate and H = Impact Height (the height from which the substrate was dropped on to the molten metal).

8.2.5 Resistance Spot Welding

In resistance spot welding the contacting surfaces are joined in one or more spots by the heat generated by resistance to the flow of electric current through work pieces that are held together under force by electrodes (Fig. 8.21). The contacting surfaces in the region of current concentration are heated by a short time pulse of low voltage, high-amperage current to form a fused nugget of weld metal. When the flow of current ceases, the electrode force is maintained while the weld metal rapidly cools and solidifies. The electrodes are retracted after each weld. The resistance welding processes thus employ a combination of force and heat to produce a weld between the work pieces.

Lou Lou and Bardon (2001) described an experimental procedure conducted to estimate and investigate the transient thermal contact conductance between the electrodes and workpieces during resistance spot welding. The electrode tip was instrumented with several interior micro-thermocouples for measuring the

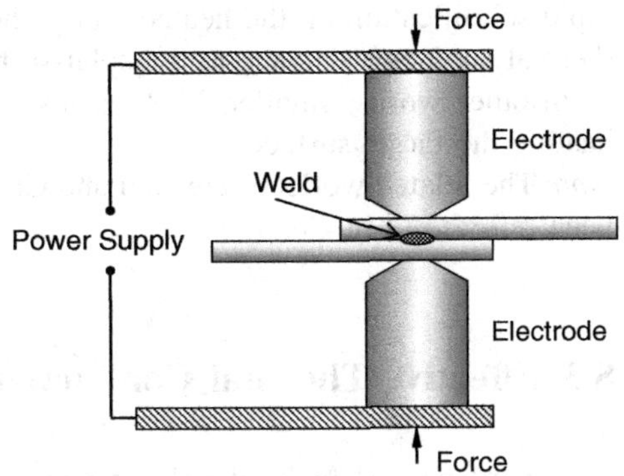

Fig. 8.21 Resistance spot welding

transient temperature response during the welding process. A simple mathematical model, using an inverse heat transfer method, was built for the estimation of the transient heat transfer coefficient from interior transient temperature measurements. A resistance welding case of two steel sheets was investigated. The initial transient values of thermal contact conductance were found to be in agreement with those observed in the dry copper–steel solid contact situation. At the end of the process, the transient heat transfer coefficient reached a high value corresponding to the maximum heat transfer rate at the interface during the welding process. When the metal is melted, the contact quality was found to increase due to the high-applied electrode force, resulting in lower thermal contact resistance

8.2.6 Thermal Spray Coating

Substrate temperature has been shown to influence the size and morphology of splats formed by molten particles impacting and solidifying during a thermal spray coating process. McDonald et al. (2007) conducted an experimental study of thermal contact resistance between plasma-sprayed particles and flat surfaces Droplets of molten zirconia, plasma sprayed onto a stainless steel substrate kept at room temperature, splashed and produced fragmented splats, whereas on a heated substrate they formed circular, disk-like, splats with almost no splashing. Since irregular splats produce porous coatings with poor adhesion strength, knowing the causes of droplet splashing is of considerable practical importance. The analysis showed that thermal contact resistance between the heated or preheated surfaces and the splats was more than an order of magnitude smaller than that on non-heated surfaces held at room temperature. Particles impacting on the heated or preheated surfaces had cooling rates that were significantly larger than those on surfaces held at room temperature, which was attributed to smaller thermal contact resistance. These observations supported the hypothesis that reduction of splat fragmentation and maximum spread diameter, due to larger cooling rates and more rapid solidification on the heated and preheated surfaces, can be attributed to low thermal contact resistance at the splat-surface interface.

In other words, smaller TCR means better coating! This is achieved by pre-heating the target surface.

Note The related works of Heichal and Chandra (2004) and Xue et al. (2007) have been referred to in Chap. 4

8.3 Effective Thermal Conductivity of Packed Beds

Stationary packed beds of solid particles are often used for storage of thermal energy, such as rock beds for solar energy and metal hydrides for the storage of hydrogen. Powder insulators, evacuated or otherwise, may also be modelled as stationary packed beds of particles. Some catalytic converters, used in the petro-chemical or automotive industries, consist of packed beds of powders. Other examples of packed beds include naturally occurring soils and rocks. In the thermal analysis of all of these applications, it is important to know the effective thermal conductivity of the packed bed. The present work deals exclusively with stationary packed beds of particles and, therefore, will be simply referred to as packed beds.

It is readily appreciated that the effective thermal conductivity of any packed bed will depend on the '*void fraction*', that is, the volume occupied by the fluid phase relative to the total volume of the bed. The void fraction is also sometimes called *volume fraction* or *porosity* and is designated by φ.

$$\emptyset = \frac{volume\ occupied\ by\ fluid\ phase}{total\ volume} \tag{8.13}$$

It should also be noted that the contact resistance between individual particles also affects the effective conductivity.

In general, conduction, natural convection, and radiation all contribute to energy transfer. Natural convection may be neglected unless very high temperature gradients and large pore sizes are involved. Radiation would be significant at high temperature especially if the conductivity of the particles is low. For example, at 1100 °C, 35 % of heat is transferred by radiation in packed beds of silicon carbide grains; this figure rises to 85 % in packed beds of glass beads (Chen and Churchill, 1963).

At low to moderate temperatures, the effective conductivity is the sum of the contact conductivity and the gas/solid conductivity. It may be noted that the model of Zehner and Schlunder (1970) assumes *point contact* between the adjacent spherical particles. Hsu et al. (1994) have recently shown that the Zehner-Schlunder model underpredicts the effective thermal conductivity of the packed sphere bed especially at high solid/fluid thermal conductivity ratios. The model based on *area contact* yields better agreement with experimental data.

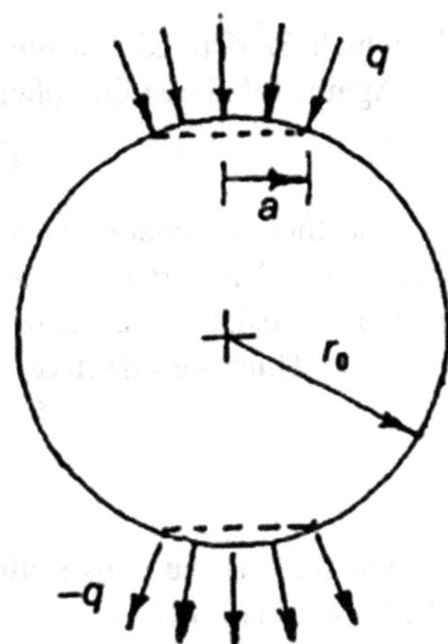

Fig. 8.22 Constriction resistance in a spherical particle

In vacuum, the heat is mainly conducted through the contact interface. The following discussion is based mainly on the analytical work of Chan and Tien (1973).

It is assumed that the radius of the contact area (Fig. 8.22) is given by the Hertzian relation

$$a = [(3/4)(1 - \nu^2)Fr_o/E]^{1/3} \tag{8.14}$$

in which F is the force between the two smooth particles.

For different packing patterns, the contact force is related to the external pressure, p, by:

$$F = S_F F_p / N_a \tag{8.15}$$

where N_a is the number of particles per *unit area* and S_F depends on the packing pattern.

For the simple cubic arrangement of spheres, there is a pair of diametrically opposite contact regions. Assuming that each contact surface is subjected to uniform heat flux, the constriction resistance associated with the solid sphere of Fig. 8.22 was shown to be

$$R_1 = 0.53/(ka), \ (a/r_o < 0.1) \tag{8.16}$$

It may be noted that this is very nearly equal to twice the disc constriction resistance, 0.27/(*ka*) , for uniform heat flux.

If the packing pattern is not simple cubic, a modified constriction resistance is defined as:

$$R = S_R R' \tag{8.17}$$

in which S_R depends on the contact radius and the wall thickness.

Again, for the solid sphere,

$$R_1' = 0.64/(ka), (a/r_o < 0.1) \tag{8.18}$$

The thermal conductance of packed spheres depends on the packing pattern. The thermal resistance of a regular packed arrangement may be considered as a group of resistances, each composed of a series of the resistances of a single particle. Thus the effective conductivity of the bed would be

$$k_e = \frac{N_a}{N_t}\left(\frac{1}{R_{ij}}\right) \tag{8.19}$$

in which R_{ij} is the constriction resistance of a single particle; subscript i refers to the type of solid ($i = 1$ for solid, $i = 2$ for hollow); subscript j refers to packing pattern ($j = 1$ for simple cubic, $j = 2$ for body centered cubic, and $j = 3$ for face centered cubic); N_a is the number of particles per unit area; N_t is the number of particles per unit length. Therefore, (N_a/N_t) has units of $[L^{-1}]$

For small (a/r_o),

$$R_{ij} = S_R S_j R_1' \tag{8.20}$$

where S_R is given in Table 8.5 for solid and hollow spheres.

The values of S_j are 1, $\frac{1}{4}$ and $\frac{1}{3}$ for j equal to 1, 2, and 3, respectively.

The expression for the conductance can then be written as

$$k_e = S_p k\left[\frac{(1-\nu^2)P}{E}\right]^{1/3} \tag{8.21}$$

where S_p depends only on the packing pattern:

$$S_p = [1.56/(S_R S_j)](N_a/N_t)(0.75 S_F r_o/N_a)^{1/3} \tag{8.22}$$

When there is no external load, the force at each contact is equal to the weight of the particles above it. The contact resistance, therefore, decreases with increasing depth from the top surface. In this case, the conductance can be shown to be

Table 8.5 Values of S_R for hollow and solid spheres (Chan and Tien 1973)

	t/r_o						
a/r_o	0.001	0.005	0.01	0.05	0.10	0.20	1 (Solid)
0.001	0.9384	0.8726	0.8479	0.8230	0.8201	0.8191	0.8252
0.002	0.9549	0.8955	0.8569	0.8071	0.8030	0.7984	0.8193
0.004	0.9582	0.9263	0.8831	0.8171	0.8489	0.8447	0.8207
0.006	0.9569	0.9380	0.9081	0.8664	0.8532	0.8236	0.8334
0.008	0.9554	0.9431	0.9192	0.8732	0.8339	0.8395	0.8280
0.010	0.9538	0.9433	0.9256	0.8588	0.8372	0.8415	0.8331

$$k_e = S_N k \left[\frac{(1 - \nu^2)\rho V L}{E r_0^2} \right]^{1/3} \tag{8.23}$$

where

$$S_N = 1.143 \left(\frac{N_a}{N_t} \right)^{2/3} \left(\frac{S_F^{1/3}}{S_j} \right) r_0^{4/3} \tag{8.24}$$

ρ is the mass density, V is the volume of the sphere, and L is the thickness of the bed. The values of the other parameters required to calculate the conductance are listed in Table 8.6.

Table 8.6 Parameters for different packing patterns (Chan and Tien 1973)

Parameter	Simple cubic	Body-centered cubic	Face-centered cubic
N_t	$1/(2r_0)$	$\sqrt{3}/(2r_0)$	$\sqrt{3}/(2\sqrt{2}r_0)$
N_a	$1/(4r_0^2)$	$3/(16r_0^2)$	$1/(2\sqrt{3}r_0^2)$
δ	0.524	0.680	0.74
S_j	1	1/4	1/3
S_F	1	$\sqrt{3/4}$	$1/\sqrt{6}$
S_p	1.36	1.96	2.72
S_N	0.452	0.713	1.02

The effect of contact resistance on the effective conductivity of packed beds was also studied by Siu (2001) and Siu and Lee (2004). Their analysis follows somewhat along the lines of Chan and Tien (1973) and use the same packing patterns - simple cubic, body centred cubic or face centred cubic. The conductivity of the medium, e.g., air, was ignored. The following features of these works are noteworthy:

The packing characteristics are usually represented by three parameters: the porosity, the mean co-ordination number and the mean contact radius. Co-ordination number is the number of contacting spheres.

Siu, approximating the mean contact area for different packing patterns, computed the effective thermal conductivity. This method relies on computing the proper constriction resistance relations and the key lies in properly accounting for the angle of contact between the contacting spheres (Fig. 8.23).

Packed beds, consisting of a single homogeneous solid, usually have a porosity of less than 0.5. These can be packing can be modelled as arrangements of spheres of uniform size (see Table 8.7).

In the above table, r_s is the radius of the spherical particle, Nc is the co-ordination number, n is the number of spheres in the unit cell and V is the volume of the unit cell. The heat transfer problem was solved numerically. Siu and Lee gave the following equations for calculating the effective thermal conductivity, ω.

$$\omega = \frac{\omega_0 R_k}{R_c + R_k} \tag{8.25}$$

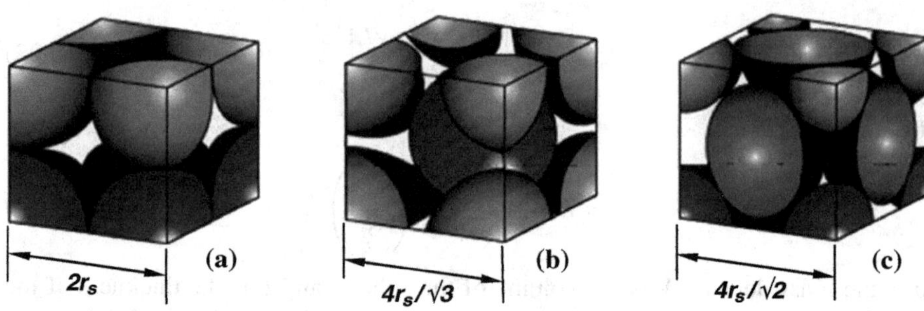

Fig. 8.23 Structure of simple unit cells: **a**: Simple cubic; **b**: Body-centred cubic; **c**: Face-centred cubic (Picture Source:chem-guide.blogspot.com; modified)

Table 8.7 Suitable packing arrangments for different porosity ranges

Packing arrangment	Porosity
Simple cubic: $N_c = 6$; $n_s = 1$; $V = 8r_s^3$	0.5–0.35
Body centred cubic: $N_c = 8$; $n_s = 2$; $V = 64r_s^3/3\sqrt{3}$	0.3–0.25
Face centred cubic: $N_c = 12$; $n_s = 4$; $V = 32r_s^3/\sqrt{2}$	<0.2

$$R_k = \frac{0.57588}{k_s r_c}\left(1 - \frac{1.0920 \times 10^{-3}}{\gamma} + \frac{3.0187 \times 10^{-5}}{\gamma^2} - \frac{1.202 \times 10^{-7}}{\gamma^3}\right) \quad (8.26)$$

$$R_c = \frac{1.75 \times 10^{-15}}{(r_c k_s)^3} \quad (8.27)$$

$$\omega_0 = \left(0.0125N_c^2 + 0.0716N_c\right)k_s\gamma \quad (8.28)$$

Note, in the above equations:

1. ω is the effective conductivity,ω_0 is the effective conductivity when the contact resistance R_c is neglected.
2. R_k is the *macroscopic* constriction resistance, taken by Siu and Lee as equal to *1/(2kr$_c$)* where r_c is the contact radius
3. γ = ratio of contact radius to radius of the sphere = r_c/r_s
4. K_s = thermal conductivity of the solid [W(mK)]
5. The units for radius is [m]
6. The numerator in the expression for R_c has units of [W^2 K^{-2}] so that R_c has units of [K/W].
7. There was a typographic error in the equation for R_c as presented in Siu and Lee (2004). This has been corrected.

The authors' results for face centred cubic arrangements showed that that the contact resistance may be significant for high conductivity materials, e.g., copper, and for low contact radii. However, the effect of R_c on effective thermal conductivity is less than 10 %, if (k_s/r_c^2) is less than 7 × 10^8. In other words, the contact resistance may be neglected if this condition is satisfied

If the gas gap conductance is significant, then it may be added to the contact conductance (see, e.g., Yovanovich 1973; Ogniewicz and Yovanovich 1977) to get the total effective conductance. The effective conductivity is obtained by multiplying the effective conductance by the thickness of the packed bed.

An experimental investigation of effective thermal conductivities of alumina-air, aluminum-air, and glass-air randomly packed beds was reported by Nasr et al. (1994). The temperature range of the tests was from 350 to 1300 K and the diameter of the particles ranged from 2.5 to 13.5 mm. The porosity in all cases was approximately 38 %. It was found that the effective thermal conductivity increased with particle size and bed temperature. At low temperatures, the heat transfer by conduction was the predominant mode; at high temperatures heat transfer by radiation became more significant. For alumina-air packed beds, the approximate values of the cross-over temperatures, that is, the temperatures at which the radiation and conduction contributions became equal are indicated in Tables 8.7 and 8.8.

Table 8.8 Temperature at which the radiation and conduction contributions became equal for alumina-air packed beds (Data from Nasr et al. 1994)

Particle diameter (mm)	Temperature (K)
2.77	1,300
6.64	860
9.61	650

The doctoral thesis of Lu (2000) was concerned with numerical modelling and experimental measurement of the thermal and mechanical properties of packed beds. His model for estimating the effective thermal conductivity is shown in Fig. 8.24

In the above diagram

G_c = solid spot conductance (Mikic)
G_g = inner gap conductance (including Smoluchowski effect)

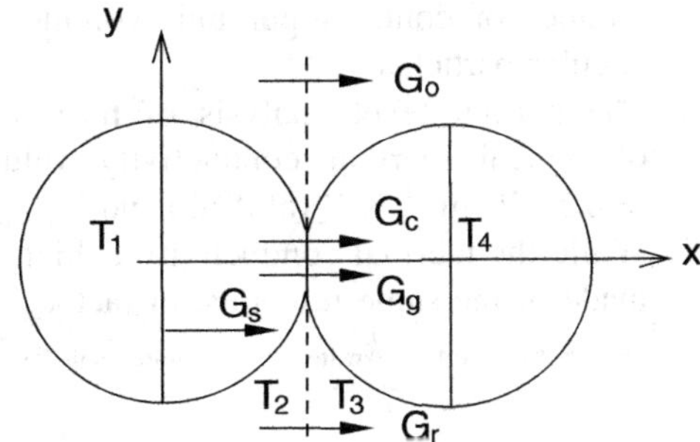

Fig. 8.24 Lu's Model identifying the heat transfer paths between contacting particles

G_s = hemispherical solid conductance
G_o = Outer gap gas conductance (calculated assuming *straight paths for the heat flow* between adjacent particles)
G_r = radiation conductance (Stefan-Boltzmann Law).

Lu's thesis is one of the few works on stationary packed beds that take into account the effect of mechanical loading. The main conclusion from Lu's thesis was that the effective thermal conductivity of metallic beds may be increased by a factor of 2–4 over a relatively small range of applied loads (~1 MPa). This was mainly due to the increase in the TCC between the particles.

Reddy and Karthikeyan (2009), considered the heat transfer due to thermal conduction (in the solid and the fluid phases of the packed bed) and the thermal contact conductance. Radiation and convection were neglected. The method used to calculate the resistance is the so-called collocated parameter model, which implies the *regular arrangement of objects side by side in a pattern*. Conduction was assumed to occur in one dimension. The geometry of the medium is considered as a matrix of touching and non-touching in-line square and circular cylinders as well as touching and non-touching in-line solid and hollow cubes. Analysis was similar to the lumped parameter model of Hsu et al. (1995). The contact resistance was not separately considered.

Reddy and Karthikeyan defined the concentration as:

$$C = \frac{volume\ of\ solid\ phase}{total\ volume} = 1 - \phi$$

A useful feature of this work is the large collection of experimental data from several previous investigators arranged in three different tables according to the concentrations: $C < 0.3$, $0.3 < C < 0.8$ and $C > 0.8$.

Yun and Santamarina (2008) measured the thermal conductivty of six different types of sand and one mixture by using a thermal needle probe (*ASTM D 5334–00*). The following conclusions are worthy of note:

1. The thermal conductivity of the dry soil linearly increases as the porosity decreases.
2. The quality of interparticle contacts and the number of contacts per unit volume govern thermal conduction in dry soils.
3. Round particles and well-graded soils tend to attain denser packing, higher number of contacts per unit volume and higher thermal conductivity than angular particles
4. The particle-level analysis of heat transfer explains the ordered sequence of typical thermal conductivity values: $k_{air} = 0.026$ W m^{-1} K^{-1}, $k_{water} = 0.56$ W m^{-1} K^{-1} (at 0 °C) and $k_{mineral} > 3$ W m^{-1} K^{-1}.
5. While the thermal conductivity is high in minerals, it is quite low in dry soils made of the same minerals, in fact $k_{dry\text{-}soil} < 0.5$ W m^{-1} K^{-1}. Thus we have: $k_{air} < k_{dry\text{-}soil} < k_{water} < k_{saturated\text{-}soil} < k_{mineral}$.

8.3.1 Correlations for Effective Thermal Conductivity

There are several experimental correlations available for predicting the effective thermal conductivity of packed beds. Listed below are three correlations chosen because: (a) all of them take the fluid as well as the solid thermal conductivities into account, and (b) they are fairly straightforward to apply. The following notation is common to all of the three correlations: λ = ratio of effective to fluid thermal conductivities; φ = void fraction or porosity; κ = ratio of solid to fluid thermal conductivities.

1. Bruggemann (1935)

$$\lambda = \frac{2}{\sqrt{3}}[(1-\kappa)\phi]^{3/2}\cos\left[\tan^{-1}\sqrt{-A}/3\right] + \kappa; 0 \le \kappa \le \kappa* \tag{8.29}$$

$$\lambda = (\kappa-1)\kappa^{1/3}\phi\left[\left\{\sqrt{A}-1/2\right\}^{1/3} - \left\{(\sqrt{A}+1)/2\right\}^{1/3}\right] + \kappa; \kappa > \kappa^* \tag{8.30}$$

where

$$A = 1 + \left(\frac{4}{27}\right)\frac{\phi^3(\kappa-1)^3}{\kappa^2} \tag{8.31}$$

and, for $0.2 \le \varphi \le 1$, κ^* is given by

$$\kappa^* = -0.00604063 + 0.139748\varphi + 0.326338\varphi^2 - 0.300101\varphi^3 + 0.901007\varphi^4(8.32) \tag{8.32}$$

Note The maximum value that κ^* can have is 1.06107, corresponding to $\varphi = 1$. For all practical solid/gas combinations, $\kappa \gg 1.06107$. Therefore, Eq. (8.29) is never applicable; Eq. (8.30) is the one to use.

2. Krupiczka (1967)

For $0.215 \le \varphi \le 0.476$,

$$\lambda = \kappa^{0.280} - 0.757^{\log(\varphi)} - 0.057^{\log(\kappa)} \tag{8.33}$$

Note

a. Krupiczka's correlation applies for cylindrical rather than spherical grains.
b. The logarithms in Eq. (8.33) are to base 10.

3. Imura and Tagekoshi 1974 (quoted in Kamiuto et al. 1989)

For $0.2 \le \varphi \le 0.476$ and $1 \le \kappa \le 10^5$,

$$\lambda = \varepsilon + \frac{1 - \varepsilon - \delta}{P + (1 - P)/k} + \kappa\delta \tag{8.34}$$

where

$$\varepsilon = \frac{\varphi - P}{1 - P}$$

and

$$P = 0.3\phi^{1.6}\kappa^{-0.044}$$

Note In this correlation, δ is an empirical factor between 0 and 0.02 accounting for contact conductance between particles. This would be of significance only for metallic powders.

Hadley (1986) derived the following semi-empirical relation for the effective conductivity, of *packed metal powders*:

$$\lambda = (1 - \alpha)\left[\frac{\phi f_0 + \kappa(1 - \phi f_0)}{1 - \phi(1 - f_0) + \kappa\phi(1 - \phi f_0)}\right] + \alpha\left[\frac{2\kappa^2(1 - \phi) + \kappa(1 + 2\phi)}{\kappa(2 + \phi) + (1 - \phi)}\right] \tag{8.35}$$

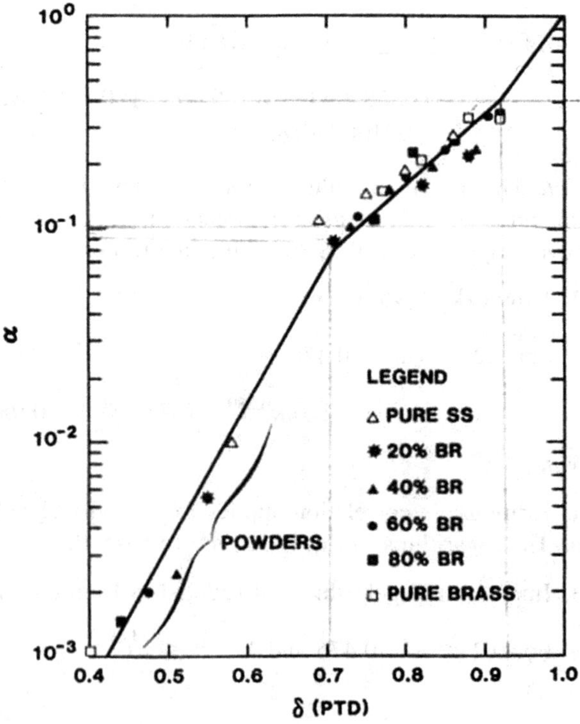

Fig. 8.25 Curve representing experimentally determined values of consolidation parameter α against percent theortical density (Hadley 1986, Reprinted by permission from Pergamon)

In this expression,

α allows for the contact between particles and depends on the degree of consolidation.

Ideally, α is determined from experiments in vacuum for randomly packed spheres. The variation of α with percent theoretical density, δ, for metallic particles, as determined by Hadley, is shown in Fig. 8.25. The percent theoretical density is the same as 'Concentration" defined earlier in this section, that is, $\delta = C = (1\text{-}\varphi)$. If the contact conductance can be neglected, then α and, therefore, the second term would be zero in Eq. (8.35). On the other hand, for evacuated powders, it is found that

$$k_{e(\text{vac})}/k = 2\alpha(1-\phi)/(2+\phi)$$

The parameter, f_o, is determined from measurements made using a high conductivity fluid, such as water. Hadley found that f_o ranged from 0.8 for (spherical) stainless steel particles to 0.9 for (angular) brass particles.

8.3.2 Use of GHP Apparatus to Measure the Effective Conductivity of Packed Beds

The guarded hot plate (GHP) apparatus (BS 874; Part 2; ASTM C177) is found to be one of the most reliable for the measurement low thermal conductivity. The essentials of a symmetrical (two-sided guarded hot plate) apparatus are shown in Fig. 8.26. Note that a standard reference material (SRM), with conductivity similar to that expected for the specimen, may be placed on one side of the hot plate for calibration purposes, if necessary. *The power inputs to the central section and the guard should be individually controlled so that the temperature of the guard matches that of the central part* (Fig. 8.27).

Madhusudana (2006a, b) conducted an experimental investigations to measure the effective thermal conductivity of packed beds of glass beads and other granular media. This work had a two-fold purpose: (a) to determine the extent of contact

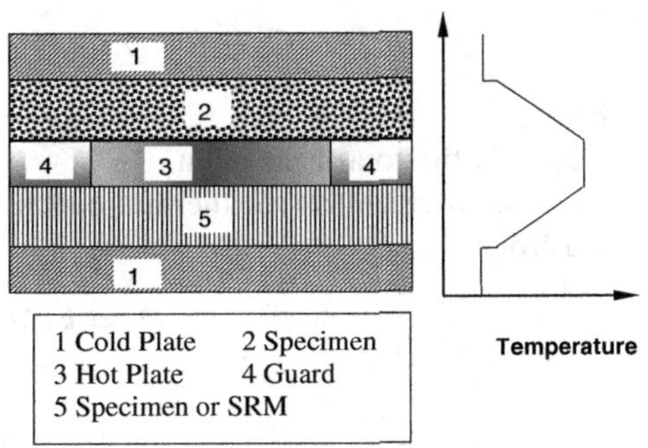

Fig. 8.26 Essentials of a symmetrical guarded hot plate apparatus

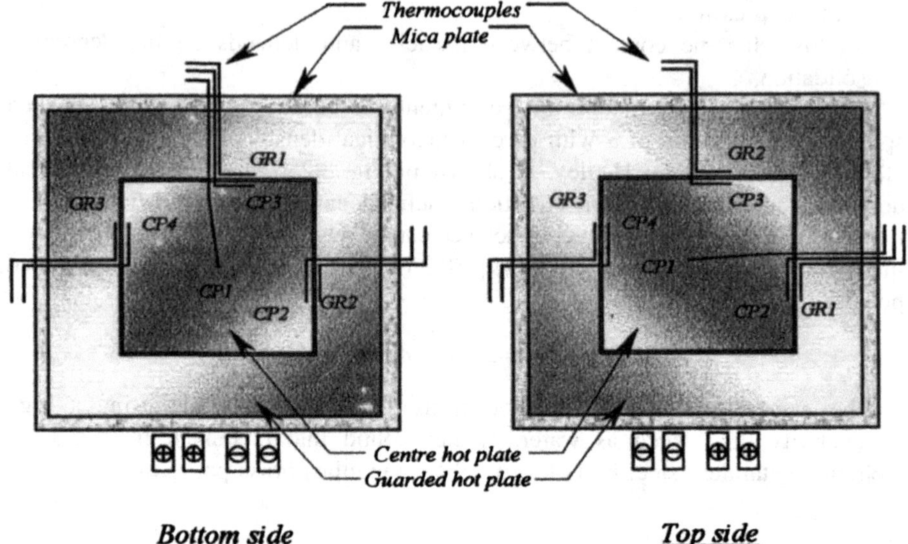

Fig. 8.27 Location of thermocouples in the hot plate and the guard ring showing the central metering section

resistances at the interfaces; and (b) the validity of the correlations listed in the previous section.

Analysis

If ΔT_R is the temperature drop in the reference specimen, then the heat flow through it is given by:

$$Q_R = k_R A(\Delta T_R/d_R) \tag{8.36}$$

in which

k_R is the thermal conductivity of the reference material
A is the area of the central metering section, and
d_R is the thickness of the reference material

Then the heat flow, Q_S, through the test specimen is obtained from:

$$Q_S = Q_{TOT} - Q_R \tag{8.37}$$

where

Q_{TOT} is the total heat input.

The *apparent* effective thermal conductivity of the packed bed is then calculated from:

$$Q_S = k_A A(\Delta T_S/d_S) \tag{8.38}$$

in which

ΔT_S is the measured temperature drop over the thickness of the test specimen, and d_S is the thickness of the test specimen.

Normally, in a GHP apparatus, ΔT_S is taken as the temperature difference between the hot plate and the cold plate. While this is satisfactory for solid materials, it is likely that significant interface resistances, R_{c1} and R_{c2}, may exist between a two-phase material such as the packed bed and its top and bottom bounding surfaces. The reason is as follows: In a solid/solid interface, the contact resistance is due to a number of microscopic constriction resistances distributed over the interface; when a packed bed consisting of several individual spheres contacts a solid surface, however, the interface resistance is due to the sum of the microscopic and the macroscopic constriction resistances

(Note that the contact resistance may be reduced by interposing a suitable material layer such as gold foil or foamed silicone rubber in the interfaces [BS 874, 1986]. It is to be emphasized, however, that this method only reduces the interface resistance, it does not eliminate it).

Thus:

$$R = R_{c1} + R_S + R_{c2} \tag{8.39}$$

where

$$R = (\Delta T_S / Q_S) \tag{8.40}$$

For this reason, it was decided to conduct the tests for two different specimen thicknesses, 25 and 50 mm, so that the interface resistances may be subtracted out as indicated below:

$$R_{25} = R_{c1} + R_{S25} + R_{c2}$$

$$R_{50} = R_{c1} + 2R_{S25} + R_{c2}$$

giving

$$R_{S25} = R_{50} - R_{25} = (\Delta T_{50}/Q_{50}) - (\Delta T_{25}/Q_{25})$$

and the actual effective thermal conductivity will be:

$$k_{E25} = h_{25}/(R_{S25}A) \tag{8.41}$$

Results for 4 mm glass beads are shown, for two different thicknesses (25 mm and 50 mm) of packed beds, in Fig. 8.28. In the chart, the thermal conductivity values are shown magnified 10 times to maintain a sense of proportion with the rest of the chart.

The void fraction was 0.369 and the average test temperature was 290 K.

The following points may be noted:

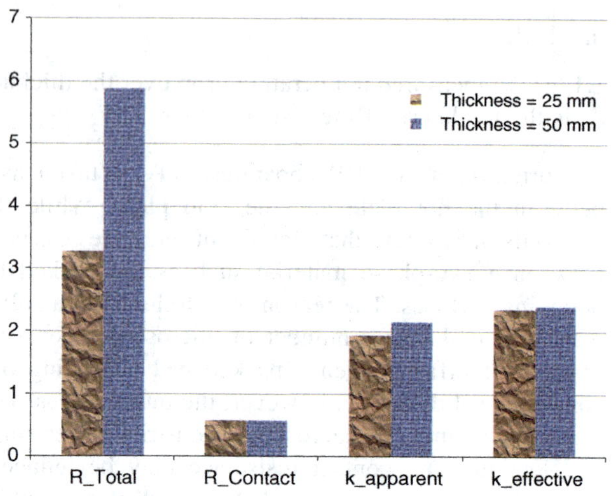

Fig. 8.28 Total resistance, contact resistance, apparent conductivity and effective conductivity for two different thicknesses of packed beds. (Note: Resistances in [m^2K/W] and thermal conductivities in [10 × W/(mK)])

1. Contact resistance is a significant part of the total resistance—nearly 18 % for bed of 25 mm thickness and about 10 % for the 50 mm thickness.
2. For the 25 mm bed, the apparent conductivity is 0.194 W/(mK) and the actual effective conductivity is 0.236 W/(mK). This means that the conductivity is underestimated by 18 % if the contact resistance is neglected. For the 50 mm bed, the apparent conductivity is 0.215 W/(mK) and the actual effective conductivity is 0.241 W/(mK). In this case, the conductivity is underestimated by 11 %.

For 6 mm glass beads (void fraction = 0.306), the effective thermal conductivity was determined to be 0.253 W/(mK) for 25 mm bed thickness and 0.259 W/(mK) for 50 mm thickness. Figure 8.29 shows the experimental results compared with the predictions using correlations of Bruggemann and Imura-Tagekoshi

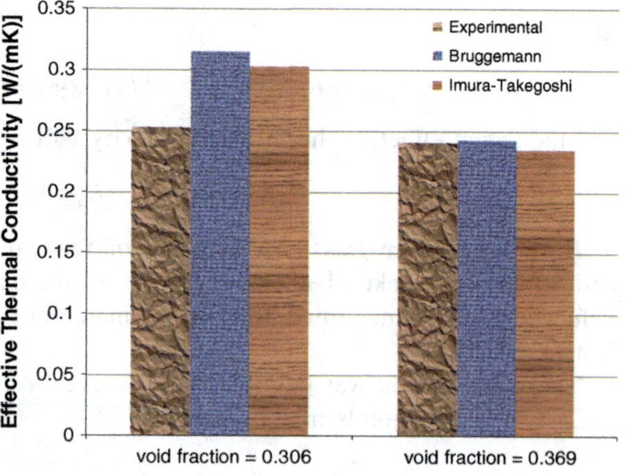

Fig. 8.29 Effective thermal conductivity of glass beads—experimental and correlation-predicted results

- It is at once seen that the smaller the void fraction, the greater is the effective thermal conductivity. This is to be expected as the thermal conductivity of air is only about (1/40)th part of the conductivity of glass.
- For the void fraction of 0.369, there is less than 1 % difference between the correlations and the experimental results.
- For the void fraction of 0.306, the correlations over predict the experimental results by about 20 %.
- Krupicska's theory is not used as the glass beads were spherical and not cylindrical. Indeed, in another series of tests (Madhusudana 2006b), it was found that Krupicska's theory matched the experimental values of effective thermal conductivity of rice grains (which may be considered cylindrical) to within 10 %.

References

Contact Heat Transfer in Finned Tubes

Cheng W (2006) Enhancing heat transfer performance of compact heat exchangers. PhD Thesis, The University of New South Wales, Australia

Cheng W, Madhusudana C (2004), Design and fabrication of a test apparatus for measuring thermal contact conductance in a finned-tube heat exchanger. Paper no. HMT-2004-C056, 17th national and 6th ISHMT-ASME heat and mass transfer conference, Kalpakkam, India

Cheng W, Madhusudana CV (2006a) Effect of soft plating on the thermal contact conductance of finned-tube heat exchangers. HVAC&R Research 12:89–109

Cheng W, Madhusudana C (2006b) Effect of electroplating on the thermal conductance of fin-tube interface. Appl Therm Eng 26:2119–2131

Dart DM (1959) Effect of fin bond on heat transfer. ASHRAE J 1:67–71

Deng J, Pagliarini G, Salvigni S (1997) A new method to estimate the thermal contact resistance in finned-tube heat exchangers. ASHRAE Trans 103:316–321

Gardner KA, Carnavos TC (1960) Thermal contact resistance in finned tubing. Trans ASME, J Heat Transfer 82:279–293

Jeong J, Kim CN, Youn B (2006). A study on the thermal contact conductance in fin-tube heat exchangers with 7 mm tube. Int J Heat and Mass Transfer, 49:1547–1555

Kulkarni MV, Young EH (1966) Bimetallic finned tubes. Chem Eng Prog 62(7):68–71

Madhusudana CV, Cheng W (2007) Decrease jn thermal contact conductance and the contact pressure of finned tube heat exchangers assembled with different size bullets. Trans ASME, J Heat Transfer 127:907–911

Piir AÉ, Roshchin SP, Vereshchagin AYu, Kuntysh VB, Minnigaleev ASh (2007) Effects of repeated high-temperature cycles on the thermal contact resistance of bimetallic finned tubes. Chem Pet Eng 43:519–522

Sheffield JW, Abu-Ebid M, Sauer HJ Jr (1985). Finned tube contact conductance: empirical correlation of thermal conductance. ASHRAE Trans 91(2a):100–117

Taborek J (1987) Bond resistance and design temperatures for high-finned tubes—a reappraisal. Heat Trans Eng 8(2):26–34

Zhao H, Salazar AJ, Sekulic DP (2009) Analysis of fin-tube joints in a compact heat exchanger. Heat Transfer Eng 30(12):931–940

Manufacturing Processes

Bendada A, Derdouri A, Lamontagne M, Simard Y (2003) Investigation of thermal contact resistance in injection molding using a hollow waveguide pyrometer and a two-thermocouple probe. Rev Sci Instrum 74:5282–5284

Bordival M, Schmidt FM, Le Maoult Y, Coment E (2007) Measurement of thermal contact resistance between the mold and the polymer for rhe stretch-blow molding process. In: Cueto E, Chinesta F (eds) CP907, 10th ESAFORM conference on material forming, pp 1245–1250

Hamasaiid A, Dour G, Loulou T, Dargusch MS (2010) A predictive model for the evolution of the thermal conductance at the casting–die interfaces in high pressure die casting. Int J Thermal Sci 49:365–372

Heichal Y, Chandra S (2004) Predicting thermal contact resistance between molten metal droplets and a solid surface. Trans ASME J Heat Transfer 127:1269–1275

Hong FJ, Qiu H–H (2007) Characterization of variable thermal contact resistance in rapid contact solidification utilizing novel ultrasound technique. Trans ASME, J Heat Transfer 129:1036–1045

Li H (2004) The role of the thermal contact resistance in the injection mold cooling of poly (ethylene terephthalate). Ph.D. thesis, The University of Toledo

Lou lou T, Bardon JP (2001) Estimation of thermal contact conductance during resistance spot welding. Exp Heat Transfer 14:251–264

McDonald A, Moreau C, Chandra S (2007) Thermal contact resistance between plasma-sprayed particles and flat surfaces. Int J Heat and Mass Transfer 50:1737–1749

Sridhar L (1999) Investigation of thermal contact resistance at a plastic-metal interface in injection molding. Ph.D. thesis, New Jersey Institute of Technology

Sridhar L, Yin W, Narh KA (2000) The effect of shrinkage induced interface gap on the thermal contact resistance between the mold and plastic in injection molding. J Injection Molding Technol 4:44–49

Xue M, Heichal Y, Chandra S, Mostaghimi J (2007) Modeling the impact of a molten metal droplet on a solid surface using variable interfacial thermal contact resistance. J Mater Sci 42:9–18

Yu CJ, Sunderland JE, Poli C (1990) Thermal contact resistance in injection molding. Polymer Eng Sci 30:1599–1606

Effective Thermal Conductivity of Packed Beds

Bruggeman DAG (1935) Dielectric constant and conductivity of mixtures of isotropic materials. Ann Phys, Series 5(24):636–664

Chan CK, Tien CL (1973) Conductance of packed spheres in vacuum. Trans ASME, J Heat Transfer 95:302–308

Chen JC, Churchill SW (1963) Radiant heat transfer in packed beds. Am Inst Chem Eng 9(1):35–41

Hadley GR (1986) Thermal conductivity of packed metal powders. Int J Heat Mass Transfer 29:909–920

Hsu CT, Cheng P, Wong KW (1994) Modified Zehner-Schlunder models for stagnant thermal conductivity of porous media. Int J Heat Mass Transfer 37:2751–2759

Imura S, Tagekoshi E (1974) Effect of gas pressure on the effective thermal conductivity of packed beds. Heat Transfer: Japanese Res 3:13

Kamiuto K, Nagumo Y, Iwamoto M (1989) Mean effective thermal conductivity of packed-sphere systems. Appl Energy 34:213–221

Krupiczka R (1967) Analysis of thermal conductivity in granular materials. Int Chem Eng 7:122–144

Lu Z (2000) Numerical modeling and experimental measurement of thermal and mechanical proerties of packed beds. Ph.D. thesis, University of California, Los Angeles

Madhusudana CV (2006a) Experimental determination of the effective thermal conductivity of packed glass beads. Paper No. HMT-2006-C152, 18th national and 7th ISHMT-ASME heat and mass transfer conference, Guwahati, India

Madhusudana CV (2006b) Low thermal conductivity measurements with a GHP apparatus. Paper No Exp-15, 13th international heat transfer conference, Sydney, Australia

Nasr K, Viskanta R, Ramadhyani S (1994) An experimental evaluation of the effective thermal conductivity of packed beds at high temperatures. Trans ASME, J Heat Transfer 116:829–837

Ogniewicz Y, Yovanovich MM (1977) Effective thermal conductivity of regularly packed spheres: basic cell model with constriction. AIAA Paper 77–188. American Institute of Aeronautics and Astonautics, New York

Reddy KS, Karthikeyan P (2009) Estimation of effective thermal conductivity of two-phase materials using collocated parameter model. Heat Transfer Eng 30:998–1011

Siu WWM (2001) Discrete formulation using thermal resistance for conduction heat transfer analysis of sphere packings. Ph.D. thesis, The Hong Kong University of Science and Technology

Siu WWM, Lee S.H.-K (2004) Contact resistance measurement and its effect on the thermal conductivity of packed sphere systems. Trans ASME, J Heat Transfer 126:886–895

Yovanovich MM (1973) Apparent conductivity of glass microspheres from atmospheric pressure to vacuum. ASME Paper 73-HT-43, American Society of Mechanical Engineers, New York

Yun TS, Santamarina JC (2008) Fundamental study of thermal conduction in dry soils. Granular Matter 10:197–207

Zehner P, Schlunder EU (1970) Thermal conductivity of granular materials at moderate temperatures. Chem Ing Tech 42:933–941 (In German)

Chapter 9
Additional Topics

In this chapter are included several areas where thermal contact conductance is important, but each area does not fit into any of the categories of previous chapters. The topics included are, contact heat transfer at low temperatures, heat transfer across stacks of laminations, effect of oxide films, specific materials including non-metallic materials. The effect of heat flow direction on the joint conductance is also considered to see under what conditions rectification can exist. The effect of loading cycles on a joint is reviewed to determine the extent of hysteresis and its practical application.

9.1 Contact Heat Transfer at Very Low Temperatures

An early impetus to the study of contact resistance at low temperatures came from a need to design strong insulating supports to large scale Dewar vessels used for storing and transporting liquids such as oxygen, nitrogen and hydrogen (Mikesell and Scott, 1956). Their work is discussed in Sect. 9.4 on "Stacks of Laminations" of this chapter.

The experimental work of Salerno et al. (1983) was concerned with the need to provide accurate thermal models for the optimum design of cryogenic instruments such as the Infrared Astronomical Satellite (IRAS). Specifically, they investigated the behaviour of pressed contacts of OFHC copper (surface finish 0.4 μm) at liquid helium-4 temperatures. The diameters of the upper and the lower specimens were 12.7 and 10.2 mm, respectively. The thermal conductance measurement apparatus was immersed in a Dewar filled with helium 4. To obtain data below 4.2 K, the temperature of the liquid helium was reduced by evaporative cooling. A pressure controller limited the evacuation to achieve the required temperature. The temperature range was, mainly, 1.6–3.8 K and the load range was 22–670 N. Calibrated germanium resistance thermometers were used to measure the temperatures of the samples.

Over this narrow range of temperature, the thermal conductance was found to vary as T^2 and ranged from 0.01 to 0.1 [W/K] depending on the load. The conductance

C. V. Madhusudana, *Thermal Contact Conductance*,
Mechanical Engineering Series, DOI: 10.1007/978-3-319-01276-6_9,

values need to be divided by the apparent contact area (the cross-sectional area of the lower specimen) to recast them into the familiar units of [W/(m^{2}K)].

Mykhaylik et al. (2012) were also concerned with the need to quantify the thermal contact conductance of copper/copper contacts in relation to the design of cryogenic instrumentation. The diameter of the samples was 14.5 mm. The temperature range was 18–70 K. typical results are shown below for the pair with a surface roughness (CLA) of 0.2 μm (two more pairs tested with surface roughness 1.6 and 3.2 μm were also tested). Their results are shown in Fig. 9.1.

The conductance values are again shown in [W/K]. These need to be divided by the area of cross-section of the specimens to express the results in [W/(m^{2}K)].

Xu and Xu (2005) studied the effects of surface topography and interfacial temperature on the thermal contact conductance of the pressed stainless steel 304 contacts in the temperature range 125–210 K and pressure range 1–7 MPa. The surface roughness of test specimens varied from 1.5 to 17.6 μm. The reduced contact conductance at low temperatures is due to the increased hardness and the decreased thermal conductivity (Figs. 9.2, 9.3).

Because of its high thermal conductivity and excellent electrical insulating property, aluminium nitride (*AlN*) is suitable to be used in electrically insulated heat conductor in high-temperature superconducting system conduction cooled by a cryocooler. Wang et al. (2006) measured the thermal conductivity of ceramic *AlN* and the thermal contact conductance between *AlN* and oxygen-free high-conductivity (OFHC) copper in the temperature range 45–140 K.

The measurements were made in a steady state axial heat flow apparatus under vacuum conditions of 10^{-3} to 10^{-4} Pa. Their results for the thermal conductivity of *AlN* are shown in Fig. 9.4.

It is worth noting that the above results pertain to commercially available polycrystalline aluminium nitride produced by liquid phase sintering. For a single crystal of aluminium nitride, the conductivity is higher by a factor of more than

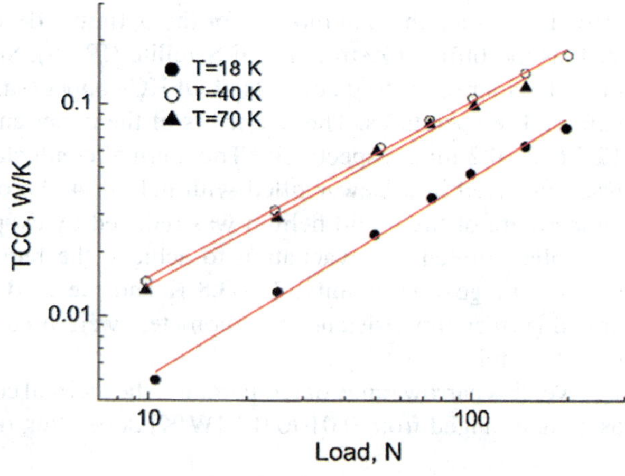

Fig. 9.1 Thermal conductance of copper/copper contacts (Mykhaylik et al. 2012. Reprinted by permission from American Institute of Physics)

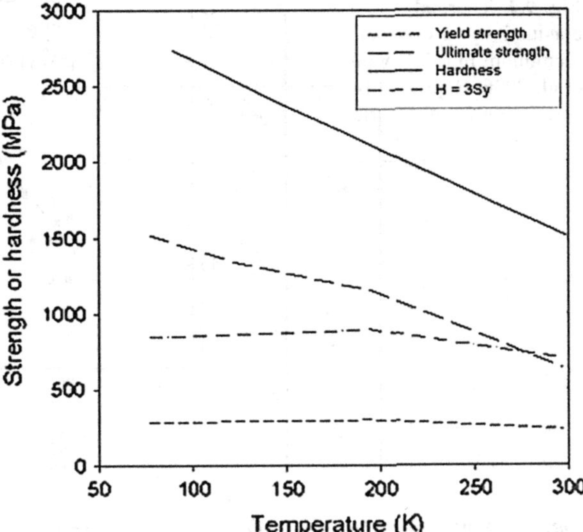

Fig. 9.2 The variation of hardness with temperature for stainless steel AISI 304 (Xu and Xu 2005; reprinted by permission from Pergamon publishers)

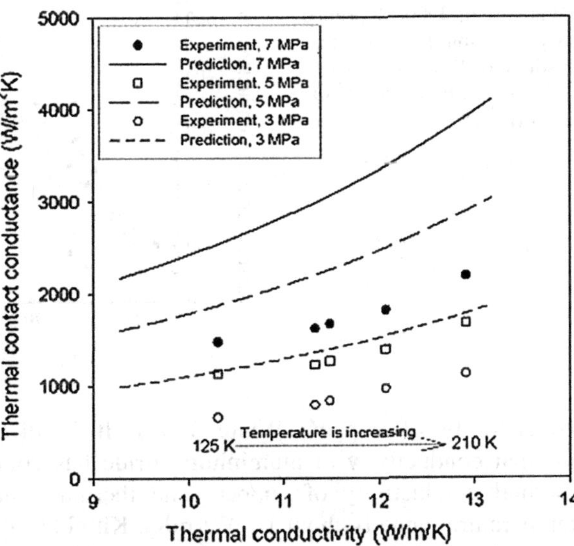

Fig. 9.3 The variation of conductance (and conductivity) with temperature for stainless steel AISI 304 (Xu and Xu 2005; reprinted by permission from Pergamon publishers)

ten, indicating that thermal boundary resistances play an important part in decreasing the effective thermal conductivity.

Their results for thermal contact conductance of aluminium nitride with OFHC copper and with Bi2223 (a high temperature superconductor) are shown in Figs. 9.5 and 9.6.

It is seen that, in both cases, the TCC increases with temperature. In the temperature range of the experiments, there is a significant decrease in the thermal

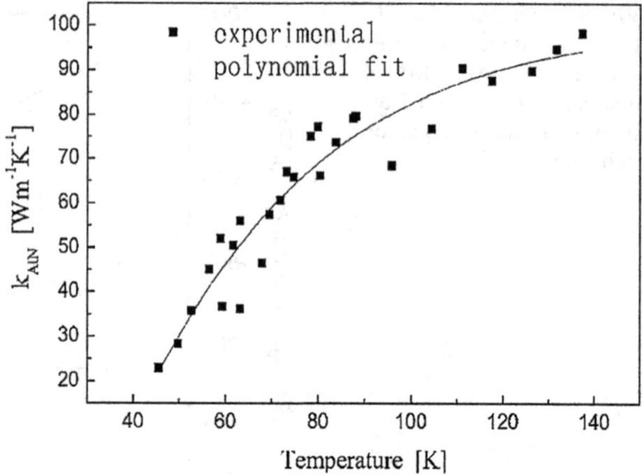

Fig. 9.4 Thermal conductivity of commercial aluminium nitride (Wang et al. 2006. Reprinted by permission from American Institute of Physics)

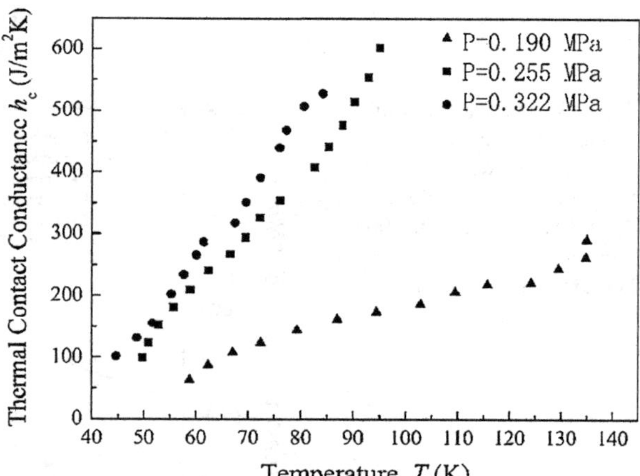

Fig. 9.5 Variation of TCC with temperature : AlN and OFHC copper contacts (Wang et al. 2006. Reprinted by permission from American Institute of Physics). *Note* TCC should be in W/(m^{2}K), not J/(m^{2}K)

conductivity of copper (White 1953). It is likely that the increase in the thermal conductivity of aluminium nitride has compensated for the decrease in thermal conductivity of copper. The thermal conductivity of Bi2223 in this temperature range is about 1.2 W/(mK), Kittel (1995). This explains the generally low values of TCC obtained for AlN/Bi2223 contacts. The change in contact pressure is quite small (0.190–0.322 MPa) for the copper/AlN contacts. Therefore it is surprising that the TCC has increased by a factor of six over this small increase in pressure.

Another experimental investigation dealing with low temperatures is the study of thermal contact conductance of molybdenum-sulphide-coated joints by Ramamurthi et al. (2007). This work has already been discussed in the Chapter on Control of TCC.

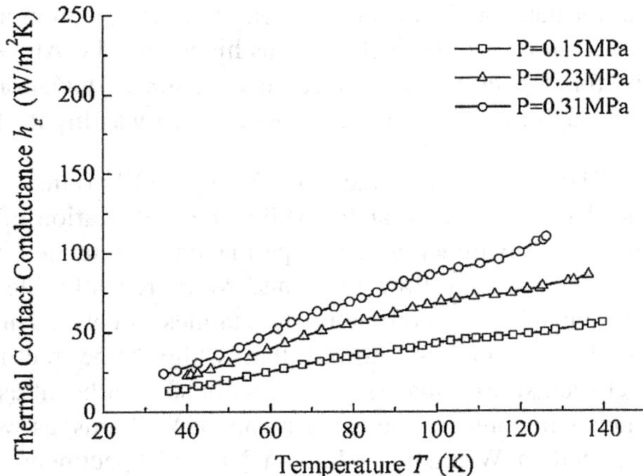

Fig. 9.6 Variation of TCC with temperature: AlN and Bi2223 contacts (Wang et al. 2006. Reprinted by permission from American Institute of Physics)

9.2 Rectification

In Chap. 6 we saw that, for radial heat flow in cylindrical joints, the magnitude of the contact conductance depends on the direction of the flow, that is, whether from aluminum to stainless steel or vice versa. This dependence of conductance on the heat flow direction, which may be called rectification, has been found to exist for apparently plane joints as well. Since the heat flows primarily through the solid spots at moderate to high contact pressures and through the gas gaps at low contact pressures, we will consider the rectification in plane joints under two separate headings.

9.2.1 Rectification at Moderate to High Contact Pressures

Barzelay et al. (1954) and Rogers (1961) were some of the earliest investigators to observe experimentally the directional effect for nominally flat specimens. In both investigations, it was noted that the conductance for heat flow from aluminum to stainless steel was higher than for the opposite direction. Clausing's (1966) tests on "spherical cap" specimens, that is, both surfaces initially convex, however, showed that the conductance was the higher when heat flowed from stainless steel to aluminum. Powell et al. (1962) measured the heat transfer coefficient at the interface between a steel ball in contact, separately, with aluminum alloy, germanium, and soapstone ceramic. No rectification effect was observed in any of the three cases. The actual contact area in these tests, however, was very small $\approx$5.5 (10^{-3}) mm^2. Hence, Powell et al. suggested that the direction effect was significant only when the contact area is large. It was also postulated that the rectification could be due to the distortion of the contact surface because of local temperature gradients. The tests of Lewis and Perkins (1968) showed that:

a. for flat, rough specimens (both flatness deviation and roughness of the order of 30 μm), the conductance was higher for the Al → SS direction.
b. for "spherical cap" specimens (flatness deviation 200–300 μm compared to roughness of 5 μm), the conductance was higher for the SS → Al direction.

The experimental studies of Williams (1976) indicated that the contact elements need not be dissimilar to exhibit the rectification effect. He also found that the rectification effect was not permanent but tended to decrease with increasing number of heat flow reversals. More recently Stevenson et al. (1989) tested combinations of nominally flat stainless steel and nickel specimens in vacuum. Their experiments showed that while some rectification existed for contact between similar materials, it was not so significant as that exhibited by dissimilar metals in contact. The experiments of Williams, as well as those of Madhusudana (quoted in Williams 1976), on Nilo 36 specimens showed that no rectification existed in this case. Since Nilo 36 is an alloy of very low thermal expansion coefficient, this confirms that thermal distortion of contacting surfaces is necessary for rectification to be present.

At present, there appears to be no single theory that can explain the rectification observed under all different conditions. However, on the basis of thermal distortions, explanations may be provided for some experimental observations.

9.2.2 Specimens with Spherical Caps

These typify plane ended specimens whose flatness deviations are large compared to the surface roughness. The theoretical model, first proposed by Clausing, is shown in Fig. 9.7. It is seen that the area of contact would be larger when heat flows from a material of high (α/k) to one of low (α/k). Thus the contact conductance for stainless steel to aluminum would be higher than for the opposite direction, confirming the experimental results for this type of contact.

We will prove the result for the spherical cap model in a more general form as follows (a somewhat different analysis was presented by Thomas and Probert 1970):

Let r_1 and r_2 be the radii of curvature of the two surfaces in contact. Then for a given normal load, F, the Hertzian contact radius, a, is (Timoshenko and Goodier 1970):

$$\begin{aligned} a &= \left[(3/4)F\left\{\left(1-\nu_1^2\right)/E_1 + \left(1-\nu_2^2\right)/E_2\right\}\{r_1 r_2/(r_1+r_2)\}\right]^{1/3} \\ &= C\{r_1 r_2/(r_1+r_2)\}^{1/3} = C\{(1/r_1)+(1/r_2)\}^{-1/3} \end{aligned}$$

in which C depends upon the mechanical load.

Since the contact conductance is proportional to the contact spot radius, a, the conductances for opposite directions of heat flow may be compared by calculating this radius for each direction and then taking the ratio of the two radii.

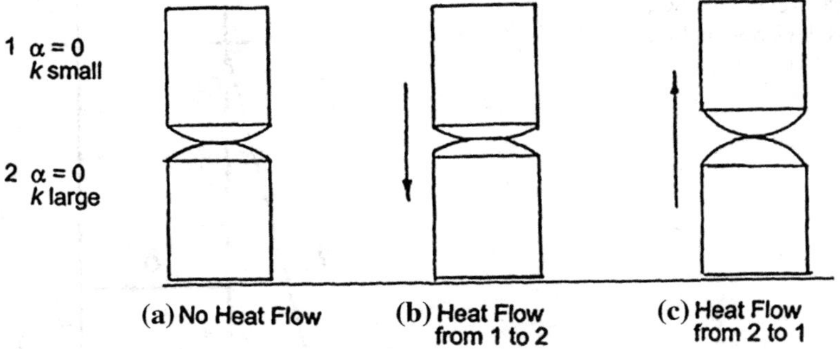

Fig. 9.7 Effect of heat flow direction in surfaces with large flatness deviation (after Clausing 1966)

If specimen 1 is heated and specimen 2 cooled, then r_1 is increased to $r_1 + \Delta r_1$ and r_2 is decreased to $r_2 - \Delta r_2$. Conversely, for the same heat flow in the opposite direction (2 → 1), r_1 is decreased to $r_1 - \Delta r_1$ and r_2 is increased to $r_2 + \Delta r_2$. In other words, when heat flows *from* the surface of a sphere, the thermal strains cause the surface to become less convex, thus increasing the radius of curvature. On the other hand, when heat flows *into* the spherical surface, the convexity is increased, thus reducing the radius of curvature. Therefore

$$(a_{12}/a_{21})^3 = \{1/(r_1 - \Delta r_1) + 1/(r_2 + \Delta r_2)\}/\{1/(r_1 + \Delta r_1) + 1/(r_2 - \Delta r_2)\}$$

Now, $1/(r_1 - \Delta r_1) = (1/r_1)(1 - \Delta r_1/r_1)^{-1} \approx (1/r_1)(1 + \Delta r_1/r_1)$, by binomial expansion. We can express the other three terms similarly. Hence, after multiplying numerator and denominator through $r_1 r_2$, we get

$$(a_{12}/a_{21})^3 = \{r_2(1 + \Delta r_1/r_1) + r_1(1 - \Delta r_2/r_2)\}/\{r_2(1 - \Delta r_1/r_1) + r_1(1 + \Delta r_2/r_2)\} \tag{9.1}$$

Thus we see that specimens made of similar materials can indeed exhibit rectification so long as $r_1 \neq r_2$. Similar conclusions were reached by the independent investigations of Somers et al. (1987). This supports the experimental observations of Williams (1976).

From geometry (Fig. 9.8), the flatness deviation is related to the radius of curvature by

$$z \approx b^2/2r$$

Hence

$$\Delta z = -(b^2/2r^2)\Delta r \tag{9.2}$$

Also

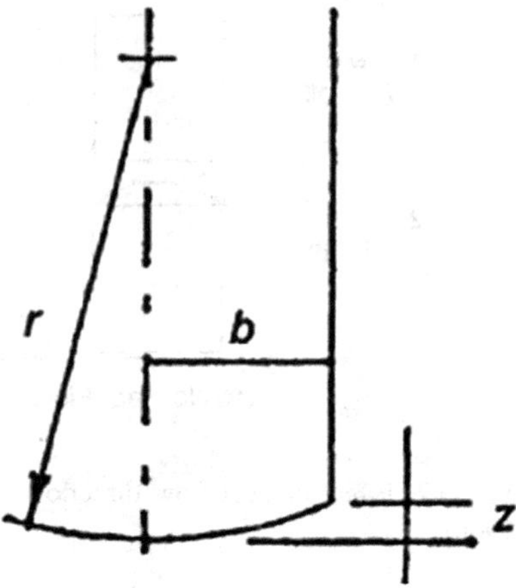

Fig. 9.8 Relation between flatness deviation and the radius of curvature

$$\Delta z = [Q\alpha(1+\nu)\ln(b/a)]/(2\pi k) \tag{9.3}$$

(see Barber 1968). In this expression, Q is negative for heat flowing away from the spherical surface, so that Δr would be positive for heat flowing from the surface, as stated earlier.

In Clausing's model, $r_1 = r_2 = r$ and α_2 is taken to be zero so that Δr_2 will be zero by virtue of Eqs. (9.2) and (9.3). Hence, from Eq. (9.1)

$$\begin{aligned}(a_{12}/a_{21})^3 &= \{r(1+\Delta r_1/r)+r\}/\{r(1-\Delta r_1/r)+r\} \\ &= (2+\Delta r_1/r)/(2-\Delta r_1/r)\end{aligned} \tag{9.4}$$

Since Δr_1 is positive, this ratio is greater than 1 and hence the conductance for the direction $1 \rightarrow 2$ will be greater than that for the opposite direction, as confirmed by the tests of Clausing (1966). Also, since the change in radius of curvature is dependent on the heat flux, we conclude that the contact conductance will be also dependent on heat flux.

9.2.3 Plane Ended Specimens

The explanation offered by Jones et al. (1974, 1975) is applicable to initially flat contacts and may be understood with reference to Fig. 9.9.

$$(\alpha_1/k_1) > (\alpha_2/k_2)$$

In a cylindrical specimen, the axial temperature gradient is constant, except in the vicinity of the disturbance represented by the contact interface. In the region of the constant temperature gradient, the two faces of a disc of radius, b, and small thickness, Δx, will suffer a relative radial displacement, $\Delta b = \varepsilon_T b = (\alpha \Delta T)b$. Because of

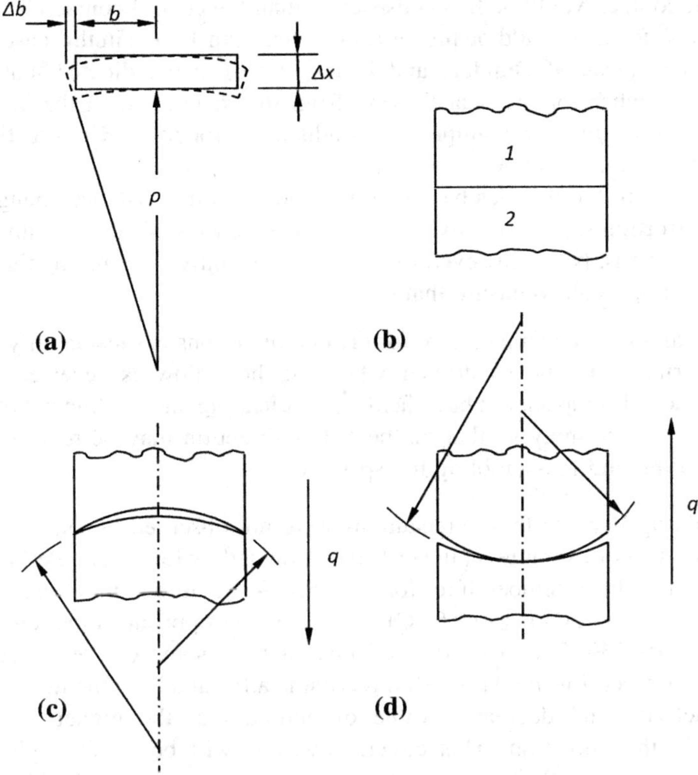

Fig. 9.9 Rectification in plane ended specimens. **a** Curvature induced due to axial temperature gradient. **b** No heat flow. **c** Heat flow from 1 to 2. **d** Heat flow from 2 to 1 (after Jones et al. 1974, 1975)

this differential radial expansion, the initially flat surfaces assume a bowed configuration, the radius curvature, ρ, of which is, by geometry, given by:

$$(\Delta b/b) = (\Delta x/\rho)$$

Hence

$$(1/\rho) = (\alpha \Delta T/\Delta x) = \alpha q/k \tag{9.5}$$

where q is the heat flux.

Thus, the radius of curvature is inversely proportional to α/k. It is assumed that, if the resulting radius of curvature is appreciably larger than the length of the specimen, then the contact face will bend to the same curvature of the elemental disc. Hence, depending on the direction of heat flow, the contact may be peripheral or central. In particular, if $\alpha_1/k_1 > \alpha_2/k_2$, and the heat flow direction is $1 \rightarrow 2$, the contact would be peripheral, whereas the contact would be central for the $2 \rightarrow 1$ direction. It can be shown further that, for a given radial contact length, the peripheral contact would be less constrictive than the central contact. Thus we see that the conductance would be higher for the direction 1 → 2 in this case also. The theoretical analysis of Dundurs and Panek (1976) also indicated that the conductance is higher for the heat flowing from the material with the higher "distortivity" $\alpha(1 + \nu)/k$. For example, the conductance for SS → Cu direction would be higher than for the opposite direction.

Thus, none of the theories based on the macroscopic resistance changes due to thermal distortion support the experimental observations of the early investigators for flat specimens. Note, however, that in experimentally determining the direction effect, it is important to ensure that:

a. The mean interface temperature for the two directions is substantially the same.
b. The surfaces are not disturbed when the heat flow is reversed; thus the experimental setup should have facilities for heating and cooling at both ends of the column assembly so that the heat flow direction may be reversed without dismantling and reassembling the specimens.

It would appear that these precautions were not observed in some of the early investigations. For example, in the tests of conducted by Lewis and Perkins (1968), the mean interface temperature for the SS → A1 direction ranged typically between 120 and 150 °F (50–65 °C), while for the opposite direction the range was from 215–250 °F (100–120 °C). Thus, at least some of the increased conductance observed for the Al → SS direction is attributable to the increased value of conductivity and decreased value of hardness at the higher temperatures obtained in this direction. This effect, however, will be small, typically 5 %, compared to the 50–100 % rectification observed by Lewis and Perkins.

9.2.4 Microscopic Resistance

The above theories consider only the macroscopic resistance. Veziroglu and Chandra (1970) observed that both microscopic and macroscopic resistances need to be taken into account in order to explain rectification, especially if the flatness deviations produced by thermal distortions are less than the order of magnitude of the surface roughness. Stevenson et al. (1989), however, found experimentally that the surface roughness had a secondary effect; it was the material properties that controlled the rectification. It has also been noted that (Somers et al. 1987) that the thermal rectification would be less if microscopic resistances are included because then the change in thermal resistance due to directional bias is a smaller percentage of the total resistance of the junction. In other words, Somers et al. attribute the directional bias to the changes in the macroscopic constriction only.

9.2.5 Rectification at Low Contact Pressures

The experimental results of Jeevanashankara et al. (1990) showed a pronounced rectification effect with the conductance for the Al → SS direction being much higher than for the SS → Al direction (see Fig. 9.10). This is contrary to the trend observed by most of the other investigators. Three relevant features of these experiments must, however, be noted:

a. The tests were conducted at low contact pressures (<0.5 MPa).
b. The tests were conducted in air, not in vacuum.
c. The mean interface temperature for the Al → SS direction was about 200 °C, whereas for the opposite direction it was about 120 °C.

Analyzing the heat flow at low contact pressures in general, Madhusudana (1993) observed first, that gas gap conductance predominates at these pressures. Second, the thermal conductivity of gases is very sensitive to temperature changes; for example, the conductivity of air increases by about 31 % over the temperature range 0–100 °C, whereas the conductivity of aluminum over the same temperature range increases by less than 2 %, and that of stainless steel by just 4 %. In view of the above factors, it was indeed to be expected that the conductance would be greater for the Al → SS direction in the tests of Jeevanashankara et al.

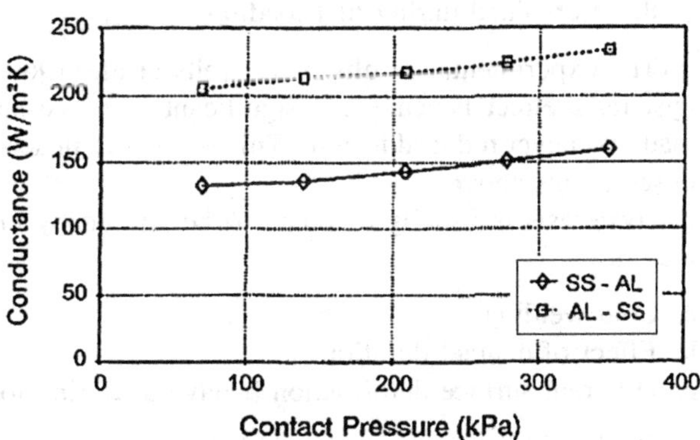

Fig. 9.10 Rectification at low contact pressure; joint in air

9.2.6 Summary

The following conclusions follow as a result of the discussion in this section.

1. At high contact pressures or for vacuum conditions, where the solid spot conduction is the predominant mode of heat transfer, the conductance would be greater when heat flows from the material of the higher distortivity, $\alpha(1 + \nu)/k$.
2. The materials need not be dissimilar to exhibit rectification effect, provided the initial convexities of the contacting surfaces are different.

3. At present, there appears to be no satisfactory explanation for the rectification observed for flat, rough surfaces in vacuum.
4. At low contact pressures when the gas gap conduction is important, rectification may be explained by means of the sensitivity of the thermal conductivity of the gas to variations in the mean junction temperature.

9.3 Hysteresis

When a joint is loaded progressively and then unloaded, the values of thermal contact conductance during unloading are often found to be higher than the corresponding values during the first loading. For example, see the experimental results of Madhusudana and Williams (1973) (Fig. 9.11 a–c) for mild steel/mild steel, zircaloy-2/zircaloy-2, and Nilo/uranium dioxide joints. In all cases a "hysteresis" loop may be seen to exist for the load–unload cycle. From the figures it may also be deduced that:

a. Hysteresis effect is more striking for coarse surfaces than for relatively smooth surfaces.
b. The hysteresis effect decreases with increasing number of cycles.
c. The conductance values eventually appear to settle down to values higher than those obtained during first loading.

The experimental results of Howells et al. (1969) had also indicated that the hysteresis effect became less significant in successive load cycles provided the load was never reduced to zero. This is roughly in accordance with two of the three observations above.

Hysteresis is usually assumed to be caused by one or more of the following factors:

a. Cold welding.
b. Effect of contact duration.
c. Different surface deformation behaviour during loading and unloading.

Cold welding requires the establishment of true metal-to-metal contact at room temperatures. For cold welding to occur, therefore, the contact must occur between perfectly clean surfaces. However, in many tests, especially those conducted in air, the surfaces are likely to be contaminated with oxide films, and so forth. Furthermore, as we have just seen, significant hysteresis has been observed for the contact between a metallic alloy (Nilo) and a ceramic (uranium dioxide). Hence it is unlikely that cold welding alone could be the cause of hysteresis.

The conductance may be expected to increase with contact duration due to the decrease of surface hardness of the specimens and hence an increase in contact area and conductance during subsequent unloading. For significant reductions in hardness to occur, the interface temperature must be well above the room

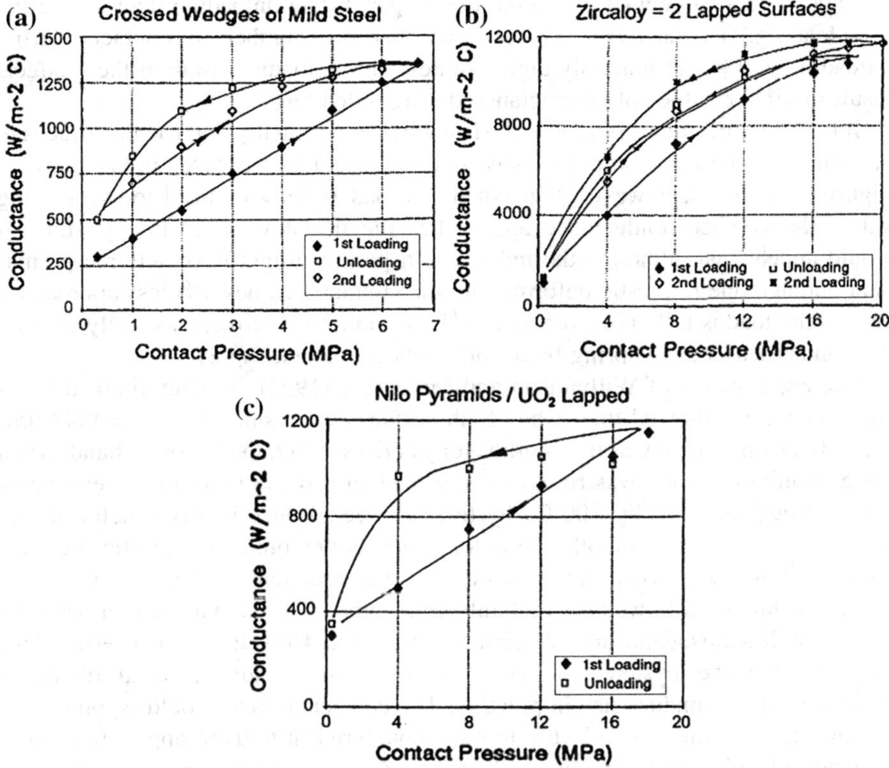

Fig. 9.11 Effect of load cycling on thermal contact conductance: **a** crossed wedges of mild steel, **b** Zircaloy—2 lapped surfaces, **c** Nilo pyramids versus UO_2 lapped surface (Madhusudana and Williams 1973)

temperature (Borzdyka 1965). Significant hysteresis effects were noticed by Fenech and Rohsenow (1963) at temperatures less than 150 °C, and by Williamson and Majumdar (1992) at temperatures well below 80 °C. At these levels, the effect of contact duration will be quite small and we can conclude that this factor will not be the main reason for hysteresis to occur.

It may thus be concluded that hysteresis is primarily due to the difference in actual areas of contact between the first and subsequent loadings of the joint. The theoretical study of Mikic (1971) assumed that during first loading the surface asperities deform plastically. For a reduction in load and subsequent increase (up to the maximum load obtained in the first loading), the asperities are expected to deform elastically, giving, as an overall effect, a higher contact area for the same contact pressure. For Gaussian distribution of asperity heights, it was found that:

a. During first loading, the actual area of contact is proportional to the pressure, p.
b. For moderate pressure reduction, $0.5 < p/p_{max} < 1$, the contact area decreases in proportion to $p^{2/3}$.

Thus the actual contact area would be larger during unloading than during the first loading. The analysis also showed that the number of contacts during unloading will be substantially higher in descending loading. Both of these effects result in higher values of conductance during unloading.

It may be noted that Mikic's analysis applies to nominally flat, rough surfaces in vacuum. The deformation of the bulk material was not considered in his analysis. Figure 9.11 shows, however, that when the load is fully reduced to the starting value, the contact conductance approaches the initial value at first loading. It would appear that, although the individual asperities might have suffered permanent damage due to plastic deformation, the contact area nevertheless approaches zero as the load is fully removed and the bulk sublayers recover elastically causing the contact surfaces to spring back and break the contact spots.

The experiments of Williamson and Majumdar (1992) on aluminium/stainless steel specimens showed that, when both surfaces were smooth ($\sigma_{\mathrm{Al}} = 0.47$ μm; $\sigma_{ss} = 0.27$ μm), there was no significant hysteresis effect. On the other hand, when the aluminium surface was rough ($\sigma_{\mathrm{Al}} = 8.71$ μm; $\sigma_{ss} = 0.40$ μm), there was a very noticeable, typically 100 %, hysteresis effect. Hence it was concluded that when the surfaces are smooth, the deformation is predominantly elastic. Furthermore, in those cases where hysteresis did exist, subsequent loading and unloading curves essentially followed the first unloading curve. This is in agreement with the results of Madhusudana and Williams (1973) and Howells et al. (1969). This suggests that the hysteresis effect may be used to advantage in obtaining enhancement of conductance in practice. The contact surface should be preloaded beyond the maximum load likely to be encountered in a given application. Subsequent unloading and reloading will yield a value of conductance, which would be higher than that which would be expected if the contact was not preloaded. A similar recommendation was made by McWaid and Marschall (1992). These observations were also confirmed by the more recent work of Li et al. (2000) on nominally flat stainsless steel/stainless steel and mild steel/mild steel joints, see Figs. 9.12 and 9.13.

Li et al. offered the following explanation for the observed hysteresis and enhancement of contact conductance due to overloading.

Consider a single asperity of tip radius R_0 in contact with a flat surface. Suppose it undergoes plastic deformation under the normal operating pressure p_0. When the load is removed, the asperity does not fully regain its original shape, but recovers to a tip radius $\mathrm{R}_1 > \mathrm{R}_0$, due to residual stresses. Provided the contact pressure is not greater than p_0 in the following load cycles, the asperity deforms elastically at the radius of R_1. The contact radius corresponding to the elastic deformation is larger than that of initial plastic deformation at the same intermediate pressure, since $a^2 = R\delta$ and R has now increased. When the asperity deforms elastically, it needs to deform further in order to sustain the same load as in the plastic deformation. Therefore, the total number of contact spots increases at the intermediate contact pressure. The combined increase in the radius and the number of contact spots leads to an increase in the TCC.

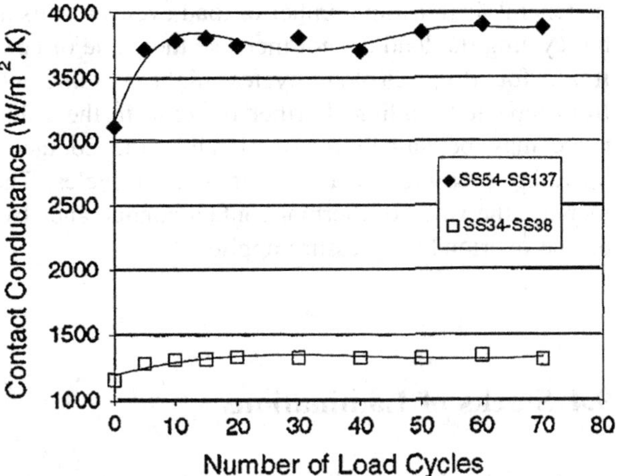

Fig. 9.12 Contact conductance versus number of loading cycles (Li et al. 2000), reproduced by permission from ASME

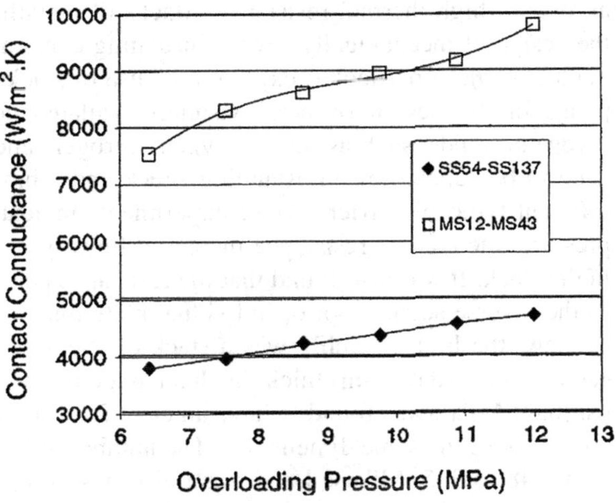

Fig. 9.13 Contact conductance versus operating pressure (Li et al. 2000), reproduced by permission from ASME

When the asperity is subjected to a new overloading pressure of $p_1 > p_0$, it deforms plastically again. It starts with an initial radius of R_1, then undergoes plastic deformation at the overloading pressure of p_1.. When the load is removed, the asperity will recover to a radius of $R_2 > R_1$. The asperity then deforms elastically in the new recovered radius of R_2 in the following load cycles, provided the pressure does not exceed p_1. As R_2 is greater than R_1, when the normal operation pressure of p_0 is applied again, the asperity will yield a larger radius of contact. The thermal contact conductance is therefore increased at the end of overloading operation even though the contact pressure applied remains the same.

Li et al. also made the following observations in the light of their experimental results:

In general, increasing number of load cycles results in higher contact conductance, but cycling the load cannot increase the value of contact conductance indefinitely. It was found that 30 load cycles might be sufficient to get the maximum benefit from repeated loading. Further increase in the value of thermal contact conductance may be gained by overloading the contact surfaces beyond the normal operating pressure for a number of load cycles. The overloading operation may increase the value of thermal contact conductance by as much as 51 %, depending on the overloading pressure applied.

9.4 Stacks of Laminations

Stacks of thin layers will provide effective thermal insulation due to the resistance to heat flow at each interface. These resistances, being in series, add up to provide an overall high thermal resistance. Stacks of metallic laminations can be used in the design of mechanically strong, insulating containers or supports. Mikesell and Scott (1956), in particular, explored the use of stacks of stainless steel and monel plates in the design of large vacuum insulated containers for the storage of cryogenic fluids, such as liquid oxygen, nitrogen, and hydrogen. These containers need to be capable of withstanding shock and vibration during transportation by different types of carriers. Their experiments indicated that, at any given contact pressure, the contact resistance increased linearly with the number of laminations in the stack. It was also found that the resistance per unit length of the stack varied as the inverse square root of individual plate thickness. From the insulating point of view, the best assembly was a stack of stainless steel plates (type AISI 302), each 0.0008″ (0.02 mm) thick. Such an assembly, when supporting a load of 1000 psi (6.89 MPa), was found to have an effective conductivity of only 2 % of a solid conductor of the same dimensions. The number of laminations in the stacks ranged from 148 to 315. Mikesell and Scott also observed that lightly dusting the plates with manganese dioxide more than doubled the resistance at all but the lowest contact pressures.

Stacks of enamelled steel laminations are used in the construction of stator cores of large turbogenerators and transformers. In this case, unlike the evacuated cryogenic supports, however, the fluid trapped between laminations is partly responsible for the heat transfer across each interface. Williams (1971), in fact, noted that the effective conductivity of such laminations is sensitive to changes of environment. The contacts are influenced more by the fluid conductivity than by the conductivity of the materials of construction. A large hysteresis was observed in the contact pressure versus strain relationship for the first cycle of loading and unloading. Subsequent loadings and unloadings showed negligible hysteresis. Williams concluded that if the generator stator core is subjected to a succession of heating and cooling cycles, while clamped tightly, there will be a tendency for the laminations to "bed down"; provided the clamping pressure is maintained, the effective conductivity should improve.

The following correlation, based on the experimental results of six different investigators, was proposed by Al-Astrabadi et al. (1977) for the conductance for stacks of thin layers in *vacuum:*

$$h^* = 3.025(P^*)^{0.58} \tag{9.6}$$

where

$h^* = h_{LL}t/k$; $P^* = P/H$
h_{LL} = layer-to-layer conductance
t = thickness of each layer
k = conductivity of the stack material
H = hardness of the stack material

The total conductance of the stack would then be given by

$$1/h = \sum(1/h_{LL}) \tag{9.2}$$

Babus'Haq et al. (1991) tested the insulating performance of multilayer ceramic (alumina) sheets in vacuum. The sheets ranged in thickness from 0.65 to 1 mm. At a contact pressure of 1 MPa, the mean thermal resistance per interface was measured as 0.55 K/W; for example, the resistance of 5 plates was 2.6 K/W and that of 15 plates was 9 K/W, approximately. They also observed that hysteresis was present in all of the tests. It was also noted that the conductance variation at low loads is due mainly to layer flattening, that is, initially the load presses out any buckles in the individual layers. At high loads, the conductance decreases mainly as a result of asperity deformation.

The effect of the mean temperature was taken into account by Fletcher et al. (1993) in the correlation of their experimental data for multilayered aluminum alloy (3004, 5042, and 5182) sheets. For thick gauge samples (sheet thickness 2.896 to 3.056 mm), the correlation was

$$h_{LL}t/k = 40{,}000(\alpha T)(P/H)^{1.15} \tag{9.7}$$

For thin gauge samples,

$$h_{LL}t/k = 2500(\alpha T)(P/H)^{0.97} \tag{9.8}$$

Since one single correlation could not be obtained to include both sets, it was suggested that consideration of another parameter, such as the surface roughness, was necessary.

9.5 Solid Spot Conductance of Specific Materials

9.5.1 Stainless Steel and Aluminium

Thomas and Probert (1972) collated 102 data points for stainless steel and 240 points for which one or both of the surfaces were aluminium alloys. These experimental results were obtained from thirteen different investigations. The correlations, together with estimates of errors, were as follows:

Stainless Steel Data

$$\ln C* = (0.743 \pm 0.067) \ln W * + (2.26 \pm 0.88) \quad (9.9)$$

Aluminum Data

$$\ln C* = (0.720 \pm 0.044) \ln W * + (0.66 \pm 0.62) \quad (9.10)$$

where

$C^* = C_s/(\sigma k)$
$W^* = W/(\sigma^2 H)$
W = Mechanical Load *(N)*
C = Thermal Conductance *(W/K)*

It is to be noted that the conductance and the load are expressed in their total values rather than in the usual, per-unit-area values, h_s and P. The range for W^* was from 10^4 to approximately 10^7, but there were comparatively small number of data points near the lower limit of the load. The relatively low values of the pressure exponents were attributed to the fact that the practical surfaces would contain some degree of waviness and would not be nominally flat. The correlation coefficients were 0.915 and 0.913 for the stainless steel and aluminum, respectively. Considering the number of different sources, these values indicate that the correlations can be called successful. No corrections, however, were applied for changes in the mean interface temperature, presumably because this information was not available in some of the experiments.

9.5.2 Zircaloy-2/Uranium Dioxide Interfaces

There is a large store of data available for this combination because of their importance to the nuclear power industry. Jacobs and Todreas (1973) considered a portion of the experimental data available and proposed a correlation whose

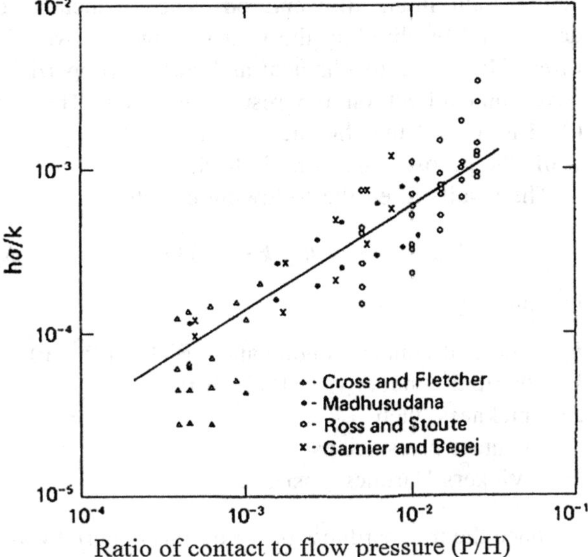

Fig. 9.14 Correlation for zircaloy-2/uranium dioxide interfaces (Madhusudana and Fletcher 1983)

constant term had to be adjusted to each set of data. Later on, Madhusudana and Fletcher (1983) collated data from five different sources[1] and proposed the correlation:

$$h_s\sigma/k = 12.29(10^{-3})(P/H)^{0.66} \tag{9.11}$$

This correlation was obtained after correcting the material properties for the different mean junction temperatures used by the various investigators. Also, in the final correlation Dean's (1962) data was omitted because, in this case, the solid spot conductances were deduced from tests in an argon environment and not direct measurements. A total of 78 data points fitted the above correlation with a correlation coefficient of 0.888 and a standard deviation of 6.78 % (the graph is shown in Fig. 9.14).

9.5.3 Porous Materials

Porous materials find application in rocket motor nozzles, combustion afterburners, heat pipes, and so on. Miller and Fletcher (1973, 1975) obtained experimental results for the contact conductance of porous copper, nickel, stainless steel, and HCR (an iron-chromium-nickel-alloy) by using these as interstitial discs between

[1] Dean (1962), Ross and Stoute (1962), Cross and Fletcher (1978), Garnier ad Begej (1979), and Madhusudana (1980).

2024-T4 aluminum alloy end rods. The conductance of the interstitial material was determined by dividing the heat flux by the overall temperature drop across the insert. Hence the mechanical and surface properties of the end rods are likely to have some effect on the results obtained. The diameter of the discs was 1″ (25.4 mm) and the thickness, t, ranged from 0.026″ to 0.091″ (0.66–2.31 mm) while the porosity, ϕ, from 30 to 86 %.

The results fitted the following correlation:

$$h_s t/k = 2.335\,[(P/H)(1-\phi)]^{0.72} \quad (9.12)$$

in which

h_s = thermal contact conductance, BTU/(h ft^2 °F)
k = thermal conductivity, BTU/(h ft °F)
t = thickness, inch
P = contact pressure, psi
H = Vickers Hardness, psi

Since the true hardness of the porous material was not known, a harmonic mean of the hardnesses of the solid interstitial material and the parent material was used. Similarly, the conductivity was also the harmonic mean of the conductivities of the parent and the solid interstitial material. The expression, as given in Eq. (9.12) and plotted in Fig. 9.15, correlated the experimental data within an average overall deviation of 16 %.

9.6 Thermal Contact Resistance in the Presence of Oxide Films

An oxide is, in general, harder (and less ductile) than its parent metal. The thermal conductivity of the oxide layer is also smaller by one or two orders of magnitude compared to that of the parent metal. For example, the hardness of mild steel EN3B is 1.67 GPa; that of iron oxide is 3.51 GPa. The thermal conductivity of

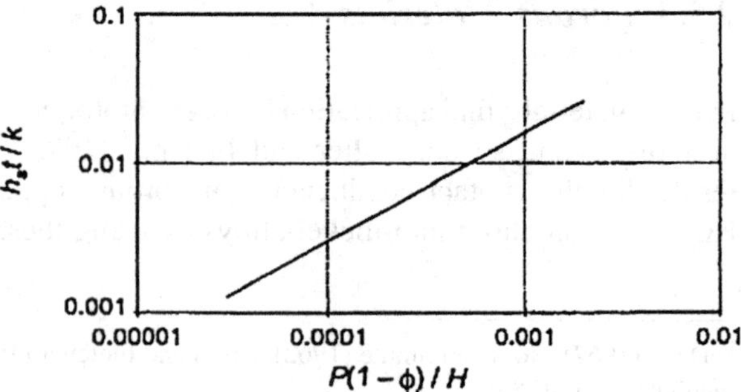

Fig. 9.15 Correlation for porous materials (based on the data of Miller and Fletcher 1975)

EN3B is 47 W/m K; that of the iron oxide is 0.875 W/m K. Thus the formation of the oxide layer tends to reduce the TCC. However, the thickness of the oxide layer on metals in seldom more than 1 μm (Kharitonov et al. 1974). Indeed, Mian et al. (1979) listed twelve different types, depending on the time duration and, hence, the colour of oxide film, of oxide layers on iron, and noted that the maximum thickness of the oxide layer ($Fe_2O_3 + Fe_3O_4$) was 0.118 μm. Hence, it is unlikely that the oxide films would have any effect on the TCC of practical surfaces which are non-flat and wavy.

The oxide films, however, might significantly affect the TCC of joints formed by *flat* surfaces. The experimental investigation of Mian et al. for oxide covered mild steel surfaces yielded the correlation:

$$R = 66\,P^{-0.945}\,\sigma^{-0.128}\,X^{0.0346} \tag{9.13}$$

in which

R = thermal contact resistance, m^2 K/W
P = contact pressure, Pa
σ = rms roughness, m
x = film thickness on each of the two contacting surfaces, m
X = total film thickness = $x_1 + x_2$.

Al-Astrabadi et al. (1980) developed a theory for oxide film covered flat surfaces based on several assumptions, including:

1. The surface heights are Gaussian and remain Gaussian when covered with the oxide layer.
2. Oxide film thickness is uniform.
3. Oxide films are thin and brittle so that metal-to-metal contact spots are formed at the mating asperities.
4. The micro contacts are of two types:
   a. Metal-to-metal surrounded by annular area of oxide.
   b. Oxide-to-oxide.

In calculating the resistance, R_1, of the metal-to-metal contact spots surrounded by oxide annuli, it was noted that the metal and oxide thermal circuits were in parallel and the effective conductivity was taken as the arithmetic mean. On the other hand, for calculating the oxide-to-oxide resistance, R_2, the metal and oxide were in series and the effective conductivity was taken as the harmonic mean. The total resistance was then

$$R_{tot}^{-1} = R_1^{-1} + R_2^{-1}.$$

It may be noted that this theory assumes that there is always either a metal-to-metal or an oxide-to-oxide contact; no gaps are postulated.

9.7 Non–Metallic Materials

The theoretical and experimental studies of Parihar (1997), Parihar and Wright (1997, 1999) focused on SS AISI304/silicone rubber/SS304 joints.

Silicone rubber is commonly used as a gasket and sealing material; it has low thermal conductivity, low hardness and is stable over a wide range of temperature (−230 to 530 K)

Existing contact resistance models, developed for interfaces of two hard materials, are not applicable to elastomer contacts because of the intrinsic properties of elastomers. For example, the stress–strain curve for an elastomer is non-linear meaning that there is no single elastic modulus. Also, elastomers are subject to thermo-mechanical softening and temperature dependent thermal conductivities and coefficients of linear expansion. The thermal conductivity of elastomers decreases with increasing temperature. Therefore, experimental measurement of contact resistance for such interfaces is needed.

It was observed that as the thickness of the specimen was decreased, or its conductivity was increased, the interface resistances played a larger role in the overall temperature drop. In other words, interface resistance of elastomer to metal contacts cannot be ignored for thin or highly conductive elastomeric layers. TCR is strongly dependent on the temperature but is a weaker function of the contact pressure.

For thick ($\delta > 3$ mm) and low conductivity elastomers, the resistances (R1 and R2) at the two (hot and cold) interfaces of a metal/elastomer/metal joints are unequal. As thickness increases, or the thermal conductivity of the elastomer decreases, the difference between R1 and R2 increases.

The differences in TCR, as well as the rectification effect, observed by the authors may be ascribed to the different thermo-physical properties existing at the hot and cold interfaces. There would be considerable temperature drop due to the low conductivity elastomer—this is exacerbated as the specimen thickness increases.

Thermal contact resistance of polymer interfaces was the subject of an experimental study by Gibbins (2006). The limited motion of polymer chains below the glass transition temperature (T*g*) allows polymers to exhibit a mechanical response much the same as metals; above T*g*, the mechanical behaviour begins to exhibit a mechanical response similar to viscous liquids and can be considered a visco-elastic material that shows a time dependent response to an applied stress (Hall 1981). For evaluations of thermal contact resistance it is necessary to avoid temperatures which bring a polymer into this viscoelastic region. Therefore only polymers which have a relatively high glassy temperature would be suitable for thermal contact resistance evaluation. Based upon these requirements, polycarbonate (Tg = 150 °C) was selected as a suitable material for the tests. It was noted, nevertheless, that the duration of load application had a noticeable effect on measured microhardness values. A reason for the difference in hardness with loading time was attributed to the possible viscoelastic behaviour of the plastic.

Heat transfer through fibrous insulation has been the subject of great interest in the aerospace community because of the use of fibrous insulation in thermal protection systems. The Lockheed Insulation LI-900 tile used in the study by Daryabeigi et al. (2012) was composed of silica fibres at a density of 141 kg/m^3. This is the most extensively characterized rigid insulation because of its use on the Space Shuttle Orbiter. The test sample used in the study had dimensions 304.8 × 304.8 mm and was 25.4 mm thick. One of the objectives of their study was to investigate whether thermal contact resistance was significant when making thermal measurements on rigid-insulation samples and, if so, to investigate a technique for eliminating it. A schematic diagram of the test setup which is somewhat similar to the guarded hot plate apparatus, is shown in Fig. 9.16.

The cold side of the rigid test sample was instrumented with six flush-mounted thin stainless steel foil thermocouples to provide the average sample cold-side temperature *TB*. Three of the foil thermocouples were coated with a thin layer of room-temperature vulcanizing (RTV) silicone, and three thermocouples were uncoated. The coated thermocouples exchanged radiant heat with the water-cooled plate more efficiently, due to their higher emittance, and therefore reached temperatures that were closer to the water-cooled-plate temperatures compared with the uncoated thermocouples.

In the absence of thermal contact resistance on the sample cold side, *TB* and *TC* should be equal. In order to eliminate thermal contact resistance on the sample cold side surface, a thin layer (1.6 mm thick) of liquid bismuth alloy was introduced between the top of the water-cooled plate and the bottom of the rigid-insulation test sample and within a containment structure. Bismuth alloy is a high thermal conductivity metal that melts at approximately 320 K. Water controlled to approximately 327 K was circulated through the water-cooled plate, thus causing the solid bismuth alloy to melt and form a liquid layer.

Significant temperature differences were measured between the sample cold-side surface and the adjoining water-cooled plate, indicating that the TCR can be considerable on rigid insulation cold side (see Fig. 9.17). Using a thin layer of liquid bismuth alloy between the sample cold-side surface and the test-setup water-cooled plate proved to be an effective method for eliminating thermal contact resistance.

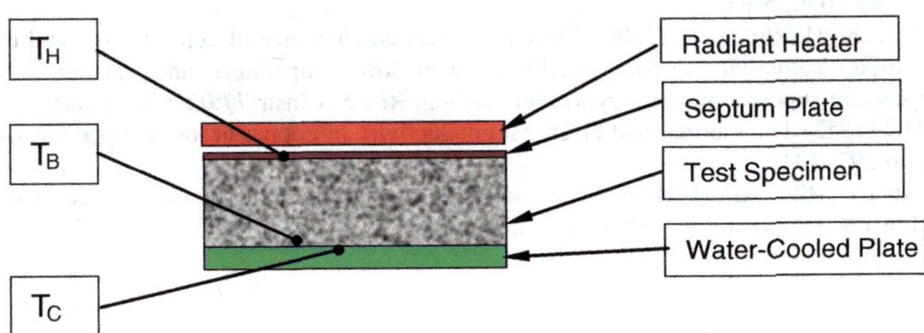

Fig. 9.16 Setup for measurement of resistance of insulation (after Daryabeigi et al. 2012)

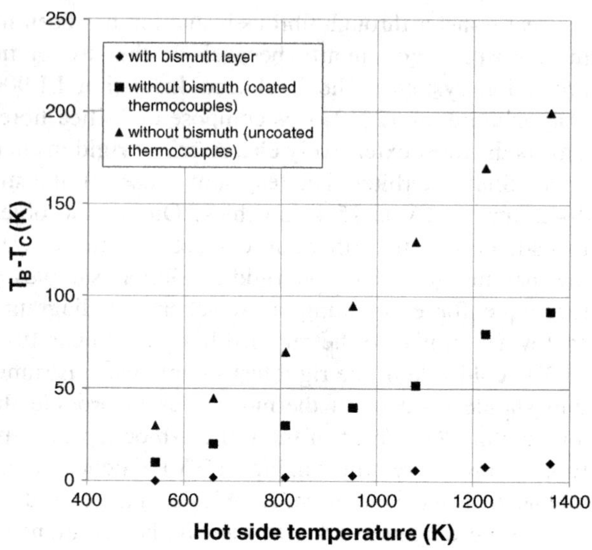

Fig. 9.17 Effect of coating thermocouples and the effect of using a thin layer of bismuth; measurements made at a vacuum of 0.001 Torr (graphs replotted based on the data of Daryabeigi et al. 2012)

References

Contact Heat Transfer at Very Low Temperatures

Kittel P (ed) (1995) Proceedings of the 1995 cryogenic engineering conference, Columbus, Ohio, p 547

Mikesell RP, Scott RB (1956) Heat conduction through insulating supports in very low temperature equipment. J Res US Natl Bur Stan 57(6):371–378

Mykhaylik VB, Burt M, Ursachi C, Wagner A (2012) Thermal contact conductance of demountable in vacuum copper–copper joint between 14 and 100 K. Rev Sci Instrum 83:034902 (6 pages)

Ramamurthi K, Sunil Kumar S, Abilash PM (2007) Thermal contact conductance of molybdenum- sulphide-coated joints at low temperature. J Thermophys Heat Transfer 21:811–813

Salerno LJ, Kittel P, Spivak AL (1983) Thermal conductance of pressed contacts at liquid helium temperatures. In: AIAA 18th Thermophysics Conference, Montreal, Canada, Paper No. AIAA-83-1436, 5 pages

Wang J, Wang H, Zhuang H (2006) Thermal contact conductance of ceramic AlN and oxygen-free high-conductivity copper interfaces under low temperature and vacuum for high-temperature superconducting cryocooler cooling, Rev Sci Instr 77:024901 (5 pages)

White GK (1953) The thermal and electrical conductivity of copper at low temperatures. Aust J Phys 6:397–404

Xu R, Xu L (2005) An experimental investigation of thermal contact conductance of stainless steel at low temperatures. Cryogenics 45:694–704

Rectification

Barber JR (1968) Comments on heat transfer at the interface of dissimilar metals—the influence of thermal strain by clausing. Int J Heat Mass Transf 11:617–618

Barzelay ME, Tong KN, Hollo C (1954) Thermal conductance of contacts in aircraft joints, US Nat Adv Comm for Aeronautics, TN 3167. National Advisory Committee for Aeronautics, Washington

Clausing AM (1966) Heat transfer at the interface of dissimilar metals—the influence of thermal strain. Int J Heat Mass Transf 9:1371–1383

Dundurs J, Panek C (1976) Heat conduction between bodies with wavy surfaces. Int J Heat Mass Transf 19:731–736

Jeevanashankara Madhusudana CV, Kulkarni MB (1990) Thermal contact conductance of metallic contacts at low loads. Appl Energy 35:151–164

Jones AM, O'Callaghan PW, Probert SD (1974) Effect of interfacial distortions on the thermal contact resistance of coaxial cylinders. In: AIAA Paper 74-689, AIAA/ASME thermophys heat transfer conference, Boston. American Institute of Aeronautics and Astronautics, New York

Jones AM, O'Callaghan PW, Probert SD (1975) Thermal rectification due to distortion induced by heat fluxes across contact between smooth surfaces. J Mech Eng Sci 17(5):252–261

Lewis DV, Perkins HC (1968) Heat transfer at the interface of stainless steel and aluminum—the influence of surface conditions on the directional effect. Int J Heat Mass Transf 11:1371–1383

Madhusudana CV (1993) Thermal contact conductance and rectification at low joint pressures. Int Comm Heat Mass Transf 20:123–132

Powell RW, Tye RP, Jolliffe BW (1962) Heat transfer at the interface of dissimilar materials: evidence of thermal comparator experiments. Int J Heat Mass Transf 5:897–902

Rogers GFC (1961) Heat transfer at the interface of dissimilar metals. Int J Heat Mass Transf 2:150–154

Somers II RR, Fletcher LS, Flack RD (1987) Explanation of thermal rectification. AIAA J 25(4):620–621

Stevenson PF, Peterson GP, Fletcher LS (1989) Thermal rectification in similar and dissimilar metal contacts. In: ASME, collected papers in heat transfer, Htd, 123:126–132

Thomas TR, Probert SD (1970) Thermal contact resistance: the directional effect and other problems. Int J Heat Mass Transf 13:789–807

Timoshenko SP, Goodier JN (1970) Theory of elasticity, 3d edn. McGraw-Hill, New York, pp. 409–413

Veziroglu TN, Chandra S (1970) Directional effect in thermal contact resistance. Heat transfer, Paper Cu 3–5, Vol. 1. Elsevier, Amsterdam

Williams A (1976) Directional effects of heat flow across metallic contacts. Inst Eng (Australia), Mech Eng Trans, Institution of Engineers (Australia), Canberra, pp 1–5

Hysteresis

Borzdyka AM (1965) Elevated temperature testing of metals. Israel program for scientific translations, Jerusalem

Fenech H, Rohsenow WM (1963) Prediction of thermal conductance of metallic surfaces in contact. Trans ASME J Heat Transf 85:15–24

Howells RIL, Probert SD, Jenkins JH (1969) Deformations of stacks of thin layers under normal compressive loads. J Strain Anal 4(2):115–120

Li YZ, Madhusudana CV, Leonardi E (2000) On the enhancement of thermal contact conductance: effect of loading history. Trans ASME J Heat Transf 122:46–49

Madhusudana CV, Williams A (1973) Heat flow through metallic contacts—the influence of cycling the contact pressure. In: First Australian conference on heat mass transfer, Sec 4.1, Monash University, Melbourne, Australia, pp 33–40

McWaid T, Marschall E (1992) Thermal contact resistance across pressed metal contacts in a vacuum environment. Int J Heat Mass Transf 35(11):2911–2920

Mikic B (1971) Analytical studies of contact of nominally flat surfaces effect of previous loading. Trans ASME J Lub Technol 20:451–456

Williamson M, Majumdar A (1992) Effect of surface deformations on contact conductance. Trans ASME J Heat Transf 114:802–810

Stacks of Laminations

Al-Astrabadi FR, O'Callaghan PW, Probert SD, Jones AM (1977) Thermal contact conductance correlation for stacks of thin layers in high vacuum. Trans ASME J Heat Transf 99:139–142

Babbus'Haq RF, Gibson C, O'Callaghan PW, Probert SD (1991) Multilayer thermally insulating ceramic contacts. J Thermophys Heat Transf 5(3): 429–434

Fletcher LS, Blanchard DG, Kinnear KP (1993) Thermal conductance of multilayered metallic sheets. J Thermophys Heat Transf 7(1):120–126

Mikesell RP, Scott RB (1956) Heat conduction through insulating supports in very low temperature equipment. J Res US Natl Bureau Stand 57(6):371–378

Williams A (1971) Heat flow across stacks of steel laminations. J Mech Eng Sci 13(3):217–223

Solid Spot Conductance of Specific Materials

Cross RW, Fletcher LS (1978) Thermal contact conductance of uranium dioxide-zircaloy interfaces. In: AIAA Paper 78–85. American Institute of Aeronautics and Astronautics, New York

Dean RA (1962) Thermal contact conductance between UO_2 and zircaloy-2. Westinghouse Electric Corporation. Report CVNA-127, Westinghouse, Pittsburgh, PA

Garnier JE, Begej S (1979) Ex-reactor determination of thermal gap conductance between uranium dioxide: Zircaloy-4 interfaces. NUREG/CR-0330, PNL-2696, Battelle-Pacific Northwest laboratories, 1979. US Nuclear Regulatory Commission, Washington, DC

Jacobs G, Todreas N (1973) Thermal contact conductance in reactor fuel elements. J Nucl Sci Eng 50:283–306

Madhusudana CV (1980) Experiments on heat flow through zircaloy-2/uranium dioxide surfaces in contact. J Nue Mater 92:345–348

Madhusudana CV, Fletcher LS (1983) Solid spot thermal conductance of zircaloy-2/uranium dioxide interfaces. J Nucl Sci Eng 83:327–332

Miller RG, Fletcher LS (1973) Thermal contact conductance of porous materials in a vacuum environment. In: AIAA Paper 73-747. American Institute of Aeronautics and Astronautics, New York

Miller RG, Fletcher LS (1975) Thermal contact conductance correlation for porous metals. Prog Astro Aero 39:81–92

Ross AM, Stoute RL (1962) Heat transfer coefficient between UO_2 and zircaloy-2, Report CRFD-1075. Chalk River, Ontario, Canada

Thomas TR, Probert SD (1972) Correlations for thermal contact conductance in vacuo. Trans ASME J Heat Transfer 94:276–281

Thermal Contact Resistance in the Presence of Oxide Films

Al-Astrabadi FR, O'Callaghan PW, Probert SD (1980) Thermal resistance of contacts: influence of oxide films. Prog Aero Astro 78:266–284
Kharitonov VV, Kokorev LS, Tyurin YA (1974) Effect of thermal conductivity of surface layer on contact thermal resistance. At Eng 36(4):308–310
Mian MN, Al-Astrabadi FR, O'Callaghan PW, Probert SD (1979) Thermal resistance of pressed contacts between steel surfaces: influence of oxide films. J Mech Eng Sci 21:159–166

Non-Metallic Materials

Daryabeigi K, Knutsonet JR, Cunnington GR (2012) Reducing thermal contact resistance for rigid-insulation thermal measurements. J Thermophys Heat Transf 26:172–175
Gibbins J (2006) Thermal contact resistance of polymer interfaces. MS Thesis, University of Waterloo
Hall C (1981) Polymer materials. The Macmillan Press Ltd., London
Parihar SK (1997) Thermal contact resistance of elastomer to metal contacts. Ph.D. thesis, University of Maryland
Parihar SK, Wright NT (1997) Thermal contact resistance at elastomer to metal interfaces. Int Comm Heat Mass Transf 24:1083–1092
Parihar SK, Wright NT (1999) Thermal contact resistance of silicone rubber to AISI 304 contacts. J Heat Transf ASME 121:700–702

Chapter 10
Concluding Remarks

Having discussed the influence of various parameters, surface configurations and types of thermal and mechanical loading, it is now possible to review the means by which the TCC (or TCR) can be controlled to suit a given practical application. The first section of the present chapter summarizes the possible methods of control; for more specific details, reference may be made to the relevant sections in the earlier chapters.

A review of the preceding chapters indicates that, over the past 50 years, a great deal of research has been carried out into the general, *fundamental* aspects of contact heat transfer. Such research includes both theoretical and experimental work. On the more *applied* aspects, however, the review shows that there is still scope for further research. Hence the second section of this chapter lists recommendations for possible future research. This list, of course, can never be complete. As new materials and processes are developed and used in equipment which need to dissipate heat, new problems are likely to arise requiring further study.

10.1 Control of Thermal Contact Conductance

10.1.1 Bare Metallic Junctions

Equation (3.33), which governs the contact conductance of nominally flat, rough surfaces may be rewritten as:

$$h\sigma/k = 1.13\ \tan\theta(P/H)^{0.94} \tag{10.1}$$

Hence, it is immediately apparent that TCC may be enhanced by:

1. Using a material combination for which the ratio, k/H, of the harmonic mean conductivity to microhardness is high.
2. Use of high contact pressures, P.
3. Use of smoother surfaces (low value of σ).

C. V. Madhusudana, *Thermal Contact Conductance*,
Mechanical Engineering Series, DOI: 10.1007/978-3-319-01276-6_10,

Conversely, high values of TCR may be obtained by using material combinations for which k/H is low or using relatively small contact pressures or rough surfaces.

In many applications, however, the choice of materials and the surface finish are dictated by other considerations such as strength, durability, corrosion resistance, and economy. Likewise, the level of contact pressure may also be specified by design considerations other than heat transfer. In such instances, we need to consider other means of control than changing the characteristics of the bare junction.

Similarly, although theoretically it is possible to control the TCC by the use of a suitable interstitial gas, in practice, one has very little control over the environment in which the heat transfer takes place. For example, in nuclear reactors the gap between the fuel and the sheath may be occupied by fission gases. For spacecraft applications, the gas gap conductance is negligible.

In dealing with bare junctions, it is also necessary to note the effect of flatness deviation. Applying Hertzian elastic contact equations to the "spherical cap" model, it can be shown that, even minor degrees of flatness deviation can result in large values of the macroscopic constriction resistance. Therefore, if thermal enhancement is the criterion, attention must be focused on getting the surfaces flat to the same degree as the roughness; it is not sufficient to simply polish the surfaces without making sure that the surfaces are also flat.

10.1.2 Use of Interstitial Materials and Coatings

The effect of interstitial materials is to fill the voids in the interface and thus provide a more continuous heat transfer path compared to that obtained in a bare junction. A joint, properly filled, is less sensitive to surface irregularities and to contact pressure variation. This means that the contact conductance is more predictable, thus contributing to reliability in thermal design. These factors make interstitial materials an attractive choice for thermal control. A very substantial amount of research, mainly motivated by the ever increasing power densities in the electronics industry, has been carried out in the past two decades on developing new thermal interstitial materials

10.1.2.1 Foils

Soft metallic foils of high thermal conductivity, for example, indium, are often used for TCC enhancement. Increases of an order of magnitude in conductance may be obtained compared to the bare junction conductance in vacuum; increases when compared to conductance in air will be somewhat less but it is possible, even in this case, to achieve significant enhancements (2 to 3 fold) by a suitable choice of the foil as indicated in Table 7.3 of Chap. 7. A very thin foil may not fill the voids completely. On the other hand, a very thick foil will result in an additional resistance in series. Thus there is usually an optimum thickness for a given surface finish beyond which the effectiveness of the foil as a conductance enhancer decreases. This optimum thickness is of the same order of magnitude as the rms surface roughness, that is, 0.48σ to 2σ, where the lower values apply for relatively harder foil materials.

In the use of foils, care should be taken to see that the foils are properly applied and remain so during operation. A wrinkled, folded, or torn foil obviously will not be as effective as one that uniformly covers the contact surface.

10.1.2.2 Metallic Coatings

Surfaces may be coated by vapor deposition, anodization, or other means. Compared to foils, coatings are more robust, uniform, and permanent. They are also likely to be more expensive. As with foils, soft metallic coatings, such as indium or tin, show maximum conductance enhancement. Again, as with foils, there exists an optimum thickness of coating for a particular surface/coating combination. The enhancement also depends on the method of deposition. For example, on aluminum alloy surfaces ($\sigma = 1{-}2$ μm), electroplated silver coating of 12.7 μm thick, yielded a thermal enhancement of about 2.5 while a flame-sprayed silver coating of the same thickness on similar surfaces yielded an enhancement of only about 0.6 to 0.7. In general, anodized aluminum coatings as well as coatings of ceramics such as silicon nitride, boron nitride, and beryllium oxide tend to decrease the TCC and hence are useful for thermal isolation. Whether it is for the purpose of conductance enhancement or isolation, both surfaces need to be coated for best results.

10.1.2.3 Greases

Thermal conductance can be enhanced by the use of various greases and lubricants. Enhancement factors of around 10 can be obtained when the bare junction is in vacuum but the figure drops to between one and two for junctions in air. Problems with greases include their tendency to run at elevated temperatures and their instability due to the loss of volatile fractions. Both of these factors tend to reduce their effectiveness as heat transfer enhancers. *Thermal Greases* are composed of a thermally conductive filler dispersed in silicone or hydrocarbon oil to form a paste.

10.1.2.4 Screens

Unless the screen material is very soft and conducting compared to the bare junction materials, wire screens are mainly used for increasing the TCR, that is, for thermal isolation. Thus, stainless steel wire screens have been found to be very effective for isolation; an order of magnitude reduction in TCC is obtainable with this type of screen. A relatively coarse mesh (small mesh number) should be used for maximum thermal isolation.

10.1.2.5 Carbon nanotubes and nano particles

Because of their theoretically high thermal conductivity along the axis, properly aligned carbon nanotubes show great promise as effective TIMs. They can be grown on a substrate and used on their own. Alternatively, they can be used to augment the effectiveness of existing TIMs such as epoxies. Although the performance of a CNT array may be improved by increasing the volume fraction, such improvement is limited because of the contact and boundary thermal resistances. Techniques for growing of CNT arrays need further improvement in order that the tubes do not buckle and all tubes make contact with the interfaces and at the right angle.

Adhesive pastes containing nanoparticles such as carbon black, and flexible graphite on which such thermal paste is applied by coating or penetrating also show great potential as TIMs.

10.1.3 Load Cycling

The conductance values upon unloading and subsequent reloadings are found to be higher than at first loading (hysteresis). Hence if a joint is preloaded to a level higher than that expected in application and the load cycled two or three times, conductance values can be expected to be higher than would otherwise be obtained. Such a load cycling may also scour the contaminant films and establish more metal-to-metal contact spots.

10.1.4 Heat Flow Direction

In many situations, the heat flow direction significantly controls the contact conductance. If a choice is possible, then for radially outward flow in cylindrical joints, heat flow should be in the direction from a material of higher *coefficient of thermal expansion* to one of a lower *coefficient* for thermal enhancement, and in the opposite direction for thermal isolation. Similarly for axial heat flow in surfaces with flatness deviation (convex surfaces) in contact, higher conductance values may be obtained by choosing the heat flow direction from the material of higher distortivity $\alpha(1+\nu)/k$ to one of lower distortivity. On the other hand, for joints at low contact pressures in a conducting medium, heat flow from the material of higher thermal conductivity to one of lower thermal conductivity will result in higher values of conductance than for the opposite direction.

10.1.5 Stacks of Laminations

When a high degree of thermal isolation as well as mechanical strength is required, stacks of thin metal laminations provide one of the best answers. Thin stainless steel laminations, several hundred in number, may have an effective conductivity which is only 2 % of that of a solid block of similar dimensions. The resistance may be further enhanced by lightly dusting the surfaces with an insulating powder, such as manganese dioxide.

10.2 Recommendations for Further Research

1. *Bolted Joints*. Theoretical results are available for both the stress distribution at the interface and the heat transfer through a bolted joint. The effect of roughness and waviness on the stress distribution needs to be further explored. Very little experimental data on heat transfer through a bolted joint appears to exist in open literature. This situation needs to be remedied.
2. *Cylindrical Joints.* The situation here is similar to that for the bolted joint. There exists some theoretical work for cylindrical surfaces without macroscopic errors of form. Future theoretical work should, therefore, include effects of waviness and out-of-roundness on the performance of a cylindrical joint. Further experimental work is also needed to substantiate the existing theories.
3. *Rectification*. Currently, there appears to be no satisfactory theory that could explain the rectification behavior observed by early investigators, namely, that the measured conductance values, in vacuum, were higher when heat flowed from a material of higher conductivity to one of lower conductivity. Also, there is no satisfactory theory explaining the rectification observed for similar materials. Research in this area should also include the effect of microscopic irregularities.
4. *Packed Beds*. The theoretical results for contact and gap conductance, obtained by assuming suitable packing patterns, need to be verified by experimental results. The theories also need to be refined. Current theoretical predictions use many simplifying assumptions: some of them neglect gas conductance, others ignore constriction resistance while only a few theories take into account the effect of mechanical loading.
5. *Microelectronics.* In view of the continuing trend toward microminiaturization of electronic circuits and consequent increased power densities, attention must be given to internal resistances in electronic packaging. This would include experimental determination of the thermal contact resistance of new mold compounds and epoxies.
6. *Manufacturing Processes.* Over the past 15 years a significant amount of experimental and theoretical research has been carried out on the role of contact conductance in processes such as die casting, injection molding and blow

forming. Other aspects of manufacturing such as chip removal and hot forging seem to have received scant attention.

7. *Heat transfer in finned tube heat exchangers.* Although it is commonly assumed that contact resistance is only of significance in heat exchangers where the tube is mechanically expanded into the fin, recent research has shown that contact resistance cannot be ignored even in tube/fin assemblies where the joint formed is a metallurgical one. It has been shown that soft electroplating of the tubes considerably enhances the performance of the exchanger; it is also enhanced by the use of larger size bullets for expansion. The limitations to these procedures need to be established. Another area of research to be explored is the decrease in contact pressure and contact conductance of the joint at operating temperatures.
8. *Periodic Contacts*. The research work on periodic contacts is surprisingly small considering its importance in internal combustion engines and several manufacturing processes. Useful information is available from some of the research work reported, but they are usually in the form graphs. Since the variables involved in periodic contact heat transfer can be related in the form of non-dimensional groups, it would be of benefit to the practising engineers if correlations between them can be derived. These should, of course, be based on reliable theoretical and experimental data.
9. *Liquid gap conductance*. Liquid metals, in general have a high surface tension. The gap resistance cannot be neglected if a liquid with high surface tension is in the interstices. At present, theories are available to predict the resistance for simple surface profiles only. There is a need to extend the theory to fit rough surfaces encountered in practice.

Erratum to: Control of Thermal Contact Conductance Using Interstitial Materials and Coatings

C. V. Madhusudana

Erratum to: Chapter 7 in: C. V. Madhusudana, *Thermal Contact Conductance*, DOI 10.1007/978-3-319-01276-6_7

In the original version of Chap. 7, the captions of Figs. 7.5 and 7.6 were interchanged. The correct captions are as follows:

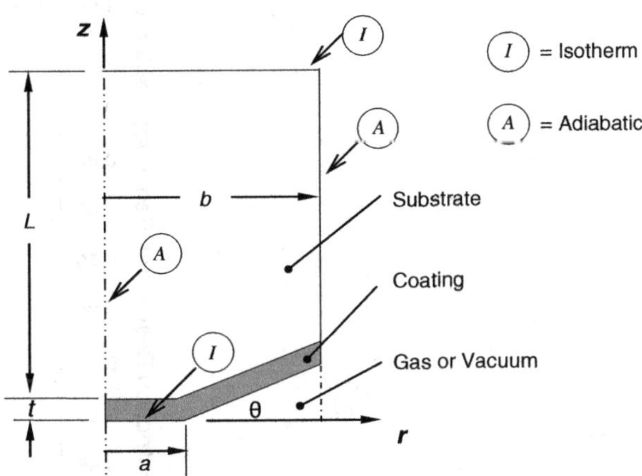

Fig. 7.5 Coated conical constriction—model for numerical analysis

The online version of the original chapter can be found at DOI 10.1007/978-3-319-01276-6_7

C. V. Madhusudana (✉)
31 Dawes, St Little Bay, Sydney, NSW 2036, Australia
e-mail: cmadhusudana@gmail.com

C. V. Madhusudana, *Thermal Contact Conductance*,
Mechanical Engineering Series, DOI: 10.1007/978-3-319-01276-6_11,

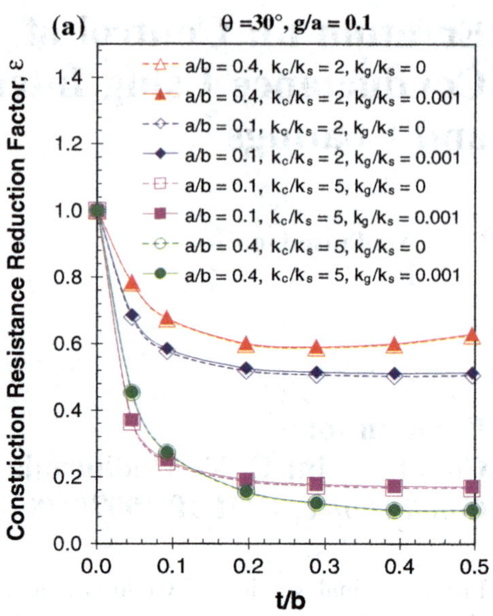

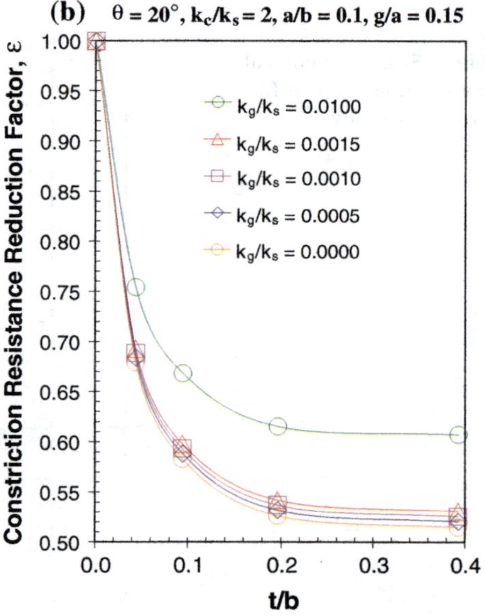

Fig. 7.6 Constriction resistance reduction versus thickness of coating. **a** Effect of varying the conductivity of the coating; **b** effect of varying the conductivity of the gaseous medium

Author Index

A
Aaron, R. L., 62
Abadi, P. P., 170
Abilash, P. M., 157, 220
Abramovitz, M., 15, 16
Abu-Ebid, M., 110, 181
Acharya, R., 60
Ajayan, Pulickel, 163
Al-Astrabadi, F. R., 159, 233, 237
Allerton, A. B., 104
Antonetti, V. W., 41, 65, 151
Aoyagi, Yasuhiro, 174
Archard, J. F., 32
Aron, W., 109
Ayers, G. H., 110

B
Babbus'Haq, R. F., 233
Bahrami, M., 41
Bairi, A., 19
Barber, J. R., 224
Bardon, J. P., 198
Barzelay, M. E., 221
Beall, A. N., 70
Beck, J. V., 18
Begej, S., 235
Bendada, A., 193, 194
Benson, H. K., 153
Biercuk, M. J., 168
Blanchard, D. G., 233
Blodgett, K. B., 60
Blum, H. A.
Boeschoten, F., 55
Bogy, D. B.
Borca-Tasciuc, Theodorian, 163
Bordival, M., 195, 196
Borisov, S. E., 58
Borzdyka, A. M., 229
Bosch, E. G. T., 86
Bowden, F. P., 128
Boysen, R. L., 177
Bradley, T. L., 100
Brown, R. E., 60
Bruggeman, D. A. G., 207, 212
Burkhard, D. G., 19
Burt, M., 218
Bush, A. W., 21, 32

C
Calkins, R. L., 161
Carnasciali, G., 144, 153
Carnavos, T. C., 181
Carslaw, H. S., 10
Cassidy, J. F., 63
Cerza, M. R., 177
Cetinkale, T. N., 20, 64, 79
Chan, C. K., 201–203
Chandra, S., 21, 74, 92, 200, 226
Chandrashekhara, K., 101
Chang, W. R.
Chao, B. T., 36
Chen, J. C., 200
Chen, P. Y. P., 21
Chen, Y. F., 106, 108
Cheng, P., 185, 187
Cheng, W., 84
Cheng, Wui-wai
Chiu C.-P., 139
Chung, D. D. L., 170
Chung, K. C., 153
Churchill, S. W., 200
Ciavarella, M.
Cividino, S., 159
Clausing, A. M., 36, 221–224

C. V. Madhusudana, *Thermal Contact Conductance*,
Mechanical Engineering Series, DOI: 10.1007/978-3-319-01276-6,

Cohen, I., 61, 82
Cola, Baratunde A., 167
Colombo, G., 109
Coment, E., 195, 196
Cook, R. S., 161
Cooper, M. G., 15, 19, 92
Costello, F. A., 140, 159
Cross, R. W., 235
Culham, J. R., 41
Cunnington, G. R., 75, 239, 240
Cunnington, G. R., Jr., 140, 160
Curti, G., 103

D
Dai, H., 162
Dargusch, M. S., 191
Dart, D. M., 181
Daryabeigi, K., 75, 239, 240
Das, A. K., 20, 21, 26
Daschyan, A. A., 111
Dean, R. A., 235
Degiovanni, A., 126
Del'vin, N. N., 61
Deng, Jinghong, 182
Derdouri, A, 193, 194
Desai, Anand Hasmukh, 167
DeVaal, J. W.
Dharmadurai, G., 58
Dodd, N. C., 84, 122
Dour, G., 191
Dryden, J. R., 148
Dundurs, J., 132, 226
Durbin, W. T., 108
Dutkiewicz, R. K., 64

E
Egorov, E. D., 111
Eichenberg, J. D, 61, 82
Elliott, D. H., 109
Etsion, I.

F
Farris, T. N., 35
Fenech, H., 229
Fernlund, I., 99, 103
Fieberg, C., 88
Field, J. S., 45
Fishenden, M., 20, 64, 79
Fisher, N. J
Fisher, Timothy S., 167
Flack, R. D., 223, 226
Fletcher, L. S., 41, 55, 79, 92, 106, 139, 142, 153, 233, 235, 236
Fried, E., 140, 159
Fullem, T. Z., 173
Furmański, P., 123

G
Ganin, Ye. A., 63
Gardner, K. A., 181
Garimella, S. V., 21, 149
Garnier, J. E., 235
Ghaffarpour. M., 125
Gibbins, Josh, 238
Gibson, C., 233
Gibson, R. D., 17, 32
Goodier, J. N., 37, 112, 222
Goodman, F. O.
Goodson, K., 162
Gould, H. H., 101
Gradshteyn, I. S., 10
Green, W. R., 62
Greenwood, J. A., 27, 31, 99
Griesinger, A., 88
Groll, E., 79, 92, 106
Gwinn, J. P., 139, 159, 161
Gyorog, D. A., 41, 79, 159

H
Hadley, G. R., 206, 208, 209
Hahne, E., 88
Hall, C., 238
Hall, R. O. A., 60
Hamasaiid A., 191
Hardee, H. C., 62
Harris, R., 56
Heichal, Y., 74, 200
Heizer, G., 159
Hirose, T., 104
Hisakado, T., 20, 27
Hohenauer, W., 86
Hollo, C., 221
Holm, R., 10
Hong, F. J., 196–198
Howard, J. R., 85, 119, 121
Howe, Timothy, 174
Howells, R. I. L., 228, 230
Hsu, C. T., 200, 206
Hsu, T. R., 82, 111
Hu, Kesong, 175
Hunter, A., 17
Hunter, A. J., 44
Hyun, J. K., 168

I

Imura, S., 208, 212
Ito, Y., 103
Iwamoto, M., 208

J

Jacobs, G., 234
Jaeger, J. C., 10, 131
Jeevanashankara, 227
Jeffrey R., 75
Jeng, D. R., 92
Jenkins J. H., 228, 230
Jeong, J., 182–184
John, J. E., 141
Johnson, A. T., 168
Johnson, R. R., 121, 122
Jolliffe, B. W., 221
Jones, A. M., 159, 224, 225
Jones, T. M., 64
Ju, Y., 35

K

Kamiuto, K., 208
Kang, T. K., 153, 156
Karplus, W. J., 92, 93
Karthikeyan, P., 206, 207
Kempers, R., 82
Kennard, E. H., 56
Kharitonov, V. V., 61, 148, 156, 237
Kim, C. N., 182–184
Kimura, Y., 27
King, M.
Kinnear, K. P., 224, 225, 233, 235, 236
Kittel, P., 220
Kneer, R., 88
Knudsen, M., 60
Knutsonet, J. R., 75, 239, 240
Koh, B., 141
Koizima, T., 103
Kokorev, L. S., 61, 148, 156, 237
Kolodner, P., 82
Krasnoborod'ko, A. I., 65, 70
Krishna Murthy, M. V., 158
Krupiczka, R., 208
Kshirsagar, B., 158
Kulkarni, M. B., 227
Kulkarni, M. V., 181
Kumano, H., 104
Kuntysh, V. B., 181

L

Lafdi, Khalid, 166
Lambert, M. A., 42, 154
Laming, L. C., 39
Lamontagne, M., 193, 194
Lang, P. M., 55
Langmuir, I., 60
Laraqi, N., 19, 26
Lasance C. J. M., 86
Le Maoult, Y., 195, 196
Lee, S. H.-K., 203, 204
Lee, S., 109
Lemczyk, T. F.
Leonardi, E., 33, 35, 44, 156, 230, 231
Leong, C.-K., 171
Lewis, D. V., 221, 226
Li, Hongbo, 193
Lin, J. T., 159, 161
Li, Y. Z., 33, 35, 44, 156, 230, 231
Linke, Heiner, 52
Litvak, A., 82
Llaguno, M. C, 168
Llewellyn-Jones, F., 10
Loulou T., 191
Loyalka, S. K., 61
Lustman, B., 61, 82
Lu, Z., 205, 206
Lyons, A., 82

M

Madhusudana, C. V., 21, 33, 35, 44, 51, 55, 82, 84, 104, 106, 110, 111, 139, 142, 151, 185, 187, 209, 213, 222, 227–230, 235
Maglic, K. D., 89
Mahajan, R., 139
Major, S. J., 21, 92
Majumdar, A., 66, 229, 230
Mal'kov, V. A., 40
Man, J. K. L., 139
Mann, D., 162
Mark, H., 63
Mark, M., 109
Marotta, E. E, 68, 144
Marschall, E., 230
Martin, D. G., 60
McDonald, Andre, 199
McIntosh, J. E, 64
McMordie, R. K., 104
McWaid, T., 230
Mian, M. N., 237

Mikesell, R. P., 217, 232
Mikic, B., 229, 230
Mikic, B. B., 15, 20, 21, 46, 79, 101, 103, 105, 109
Miller, R. G., 160, 235, 236
Milosevic, N. D., 89
Minakuchi, Y., 103
Minnigaleev ASh., 181
Mirmira, S. R., 42
Misra, M., 71
Mittlebach, M., 109
Mohs, W. F., 149
Moore, C. J.
Moran, K. P., 109
Moreau, Christian, 199
Mortimer, M. J., 60
Moses, W. M., 84, 121, 122
Mostaghimi, J., 74, 200
Motosh, N., 101
Muthanna, S. K., 101
Mykhaylik V. B., 218

N
Nagaraju, J., 71, 158
Nagumo, Y., 208
Nasr, K., 205
Negata, S., 103
Negus, K. J., 17–19, 21, 67
Nekrasov, M. I., 111
Nho, K., 68
Nishino, K., 63
Noorpoor, A. R., 125

O
O'Callaghan, P. W., 79, 142, 224, 225, 233, 237
Ochterbeck, J. M., 162
Oehler, S. A., 104, 109
Ogniewicz, Y., 204
O'Keefe, T. J., 153
Olander, D. R., 60
Olsen, E., 149
Olsen, E. L., 21

P
Pagliarini, Giorgio, 182
Pal, Sunil, 163
Panek, C., 226
Parihar, S. K., 238
Perkins, H. C., 221, 226
Persson, Ann, 42
Peterson, G. P., 92, 106, 153, 222, 226
Phelan, P. E., 50
Phillips, G. W., Jr., 81
Piir, A. É., 181
Pikus, VYu., 111
Pop, E., 162
Popov, V. M., 41, 65, 70
Powell, R. W., 221
Prasher, R., 52, 139
Probert, S. D., 79, 142, 222, 234, 237
Pullen, J.

Q
Qiu, H.-H, 196

R
Radosavljevic, M., 168
Raffa, F., 103
Rajamohan, V., 151
Ramadhyani, S., 205
Ramamurthi, K., 157, 220
Rapier, A. C., 64
Raynauld. M., 89
Razani, A., 70
Reddy, K. S., 206, 207
Remington, C. R., 159, 161
Roberts, J. K., 60
Robinson, A. J., 82
Roca, R. T., 33, 103, 105, 109
Rogers, G. F. C., 221
Rohsenow, W. M., 15, 20, 21, 27, 79, 229
Rosenfeld, A. M., 18
Roshchin, S. P., 181
Ross, A. M., 235
Rötscher, F., 97
Ryder, E. A, 17
Ryzhik, M., 10

S
Sadhal, S. S., 20, 21
Sadhal, S. S., 26
Salerno, L. J., 217
Salvigni, Sandro, 182
Samuelson, Lars, 42
Sanokawa, K., 20
Santamarina, J. C., 207
Sauer, H. J., Jr., 92, 110, 132, 159, 161, 181
Sawa, T., 104
Sayles, R. S., 37
Schlunder, E. U., 200
Schmidt, F. M., 195, 196

Schneider, G. E., 21, 89
Scott, R. B., 217, 232
Semyonov, YuG., 58, 60
Sexl, R. U., 19
Shaikh, R. A., 70
Shaikh, Shadab, 166
Sheffield, J. W., 92, 110, 153, 181
Sherbrooke, P.
Shibuya, T., 103
Shi, Li , 42
Shlykov, YuP., 63
Shojaefard, M. H., 125
Siegel, Richard, 163
Silverman, Edward, 166
Simard Y., 193, 194
Simons, R. E., 41
Sinha, N., 162
Siu, W. W. M., 203, 204
Smoluchowski, M. S., 60
Smuda, P. A., 79
Snaith, B., 139, 140
Sneddon, I. N., 45, 98
Somers, II. R. R., 223, 226
Song, S., 59, 68, 109
Son, Youngsuk, 163
Spindler, K., 88
Spivak, A. L, 217
Sridhar. L., 194, 195
Stafford, B.D.
Stegun, I. A., 15, 16
Stevenson, P. F., 222, 226
Stewart, W. E., Jr, 159
Stoute, R. L., 235
Strona, P., 103
Strong, A. B., 89
Suetin, P. E., 58, 60
Sunil Kumar, S., 157, 220
Sutton, A. E., 119, 121
Swain, M. V., 45

T
Tabor, D., 31, 32, 46, 128
Taborek, J., 182
Tagekoshi, E., 208, 212
Tam, W. K., 82, 111
Teertstra, P., 168
Thomas, L. B., 60, 61
Thomas, T. R., 37, 222, 234
Thompson, J. C., 21
Tien, C. L., 40, 189, 201–203
Timoshenko, S. P., 37, 111, 222
Timsit, R. S., 17, 73
Todreas, N., 234
Token, K. H., 161
Tong, K. N., 221
Torii, K., 63
Toyoda, J, 103
Tripp, J. H., 27
Trumpler, P. R., 58, 59
Tsukizoe, T., 20, 27
Tye, R. P., 221
Tyurin, Yu. A., 148, 156, 237

U
Ullman, A., 60
Ursachi, C., 218

V
van der Held, E. F. M., 55
Veilleux, E., 109
Venart, J. E. S, 21, 92
Vereshchagin, AYu., 181
Veziroglu, T. N., 21, 91, 92, 226
Vickerman, R. H., 56
Villanueva, E. P., 81, 142
Viskanta, R., 205
Vogd, C.

W
Wachman, H. Y., 58
Wagner, A., 218
Wahid, S. M. S., 55, 68
Wang, H., 126, 218, 220, 221
Wang, J., 218, 220, 221
Wang, Q., 162
Webb, R. L., 139, 159, 161
Weills. N. D., 17
Wen, C. Y., 108
White, G. K., 220
Whitehouse, D. J., 32
Whitehurst, C. A., 108
Whittle, T. D., 41
Wiedmann, M. L., 58
Williams, A., 21, 79, 82, 92, 111, 222, 223, 228–230, 232, 233
Williamson, J. B. P.
Williamson, M., 66
Wiśniewski, T. S., 123
Wong, K. W., 200, 206
Wood, R. A., 92, 110
Wright, N. T., 238
Wu, C. H., 108

X
Xu, Jun, 167
Xu, Lie, 218, 219
Xu, Ruiping, 218, 219
Xue, M., 74, 200

Y
Yeh, C. L., 108
Yeh, S. H., 106
Yeo, T., 133
Yeow, J. T. W., 162
Yip, F. C, 21, 92, 104
Yoshemine, K., 103
Youn, B., 182, 183
Young, E. H., 181
Yovanovich, M. M., 17, 37, 59, 67, 89, 109, 142, 151, 204
Yun, T. S., 207

Z
Zehner, P., 200
Zhan, L., 45
Zhou, Feng, 51, 52
Zhuang, H., 218, 220, 221

Subject Index

A

Actual area of contact, 1, 130, 230
Aircraft structural joints, 3
Aluminium nitride, 218–220
Analogue methods
  electrolytic tank, 92
Average clearance, 27, 30
Axial heat flow apparatus
  guard heater, 80, 82
  radiation shield, 94
  thermocouples, 80
  vacuum pump, 80

B

Bi2223, 219, 220
Bimetallic finned tubes, 182
Bolted and riveted joints, 4
Bolted or riveted joints
  effect of other parameters, 103
  heat transfer
    bolt hole radius, 105
    contact zone radius, 106
    macroscopic resistance, 106
    plate thickness, 105
  stress distribution, 97
  summary, 110
  transverse heat flow, 109

C

Carbon nanotubes, 53, 139, 162–164
  summary, 168
Coating material
  beryllium oxide, 154, 247
  copper, 141, 142, 153
  gold, 142
  indium, 140, 141, 153, 160, 166
  lead, 140–142, 153
  molybdenum sulphide, 156
  silicon nitride, 158
  silver, 144, 151, 153
Coatings
  electroplating, 139, 151
  vacuum deposition, 139
Constriction alleviation factor, 17, 18, 25
Constriction resistance
  bounded by a semi-infinite cylinder, 14
  constant temperature, 15
  uniform heat flux, 18
  circular disc in half space, 9
    constant temperature, 11
    uniform heat flux, 11
  in a fluid environment, 19
  other types, 21
Constriction resistance of plated contact, 149
Contact heat transfer
  mechanism, 1
  significance, 3
Contact heat transfer at very low temperatures, 217
Contact heat transfer in finned tubes, 181
Contacts spots
  average size, 25, 27
  number, 25, 27, 29, 31, 34, 39
Control
  bare metallic junctions, 245
  carbon nanotubes and nano particles, 248
  foils, 246
  greases, 247
  heat flow direction, 248
  load cycling, 248
  metallic coatings, 247
  screens, 247

C. V. Madhusudana, *Thermal Contact Conductance*,
Mechanical Engineering Series, DOI: 10.1007/978-3-319-01276-6,

Control (*cont.*)
stacks of laminations, 249
use of interstitial materials and coatings, 246
Convection, 55
Cooling in electronic systems, 3
Cryogenic instrumentation, 218
Cryogenic liquids, 4
Cylindrical joints
differential expansions, 111, 113
numerical example, 115
pressure, 111, 115–119

D
Deformation analysis, 25, 31
Die casting, 190, 191

E
Eccentric constrictions, 19
Epoxies
Eccobond, 173
silver filled, 175
Experimental aspects
accuracy, 79, 93
summary, 95

F
Finned tube heat exchanger electroplated tubes
gold, 186
silver, 186
tin, 186
zinc, 186
hoop stresses, 189, 190
pressure, 187, 189, 190
variation with temperature, 189
variation with size of expansion bullet, 190
solid spot conductance, 187, 188
variation with temperature, 188
variation with size of expansion bullet
Finned tubes
electroplated tubes, 186
examples, 182
maximum recommended bond temperatures, 182
thermal contact conductance, 184
correlation for 7 mm dia tubes, 184
hydrophilic coating, 184
Foils
aluminium, 145
copper, 145
gold, 142
indium, 145
lead, 145
tin, 142

G
Gap conductance
conclusions, 72
correlations, 63, 68
effect gas pressure, 62
factors affecting, 55
Knudsen number, 57
mean free path, 56, 61, 62
mean thickness of gas gap, 55
numerical example, 66, 67
Prandtl number, 56
ratio of specific heats, 56
recent research, 68
Smoluchowski effect, 56
temperature jump
mixture of gases, 56
thermal accommodation coefficient
effect of temperature, 60
summary, 62
Gap fluid conductance
defined, 3
Gap fluid is liquid
density, 73, 75
surface tension, 72, 73, 75
Gap thickness, 1
Gaussian distribution of heights and slopes, 30
Greases, 139

H
Hysteresis
crossed wedges, 229
lapped surfaces, 229
pyramids/lapped, 229
cold welding, 228
contact duration, 228, 229
number of loading cycles, 231
overloading, 231

I
Injection moulding
bond line thickness (BLT), 169
flexible graphite, 171, 176
Cabot coated, 175
Cabot penetrated, 175
insulating interstitial materials
felt, 159

mica, 159
teflon, 159
Interstitial materials, 139–141, 162

L
Laser-assisted direct imprinting (LADI), 4
Liquid gap conductance, 250

M
Macroscopic irregularities, 36
Manufacturing processes, 4, 181, 190

N
Non metallic materials, 238
Nuclear reactor, 3

O
OFHC copper, 217, 219
Oxide films
thermal contact resistance, 236

P
Packed beds
concentration (or percent theoretical density), 209
correlations, 207
effective conductivity, 200, 202–204, 212
unit cell, 203
structure (or packing arrangement), 204
void fraction (also porosity or volume fraction), 200, 207, 211–213
Periodic contacts
Biot number, 121
diffusivity, 125
equivalent length, 121
Fourier number, 121
quadrupole method, 126
quasi steady state, 128
summary, 128
Phase change materials (PCMs)
low melting point alloys, 161
Phonon mean free path, 49
Plasticity index, 32, 43
Powders, 139
Power spectral density, 32

R
Radial heat flow apparatus, 82, 94
Radiation, 1
Radius of macroscopic contact area, 37
Rapid contact solidification
ultrasound echoes, 197
Ratio of real to apparent area of contact
elastic deformation, 33
plastic deformation, 32
Rectification
low contact pressures, 227
microscopic resistance, 226
moderate to high contact pressures, 221
plane ended specimens, 224
specimens with spherical caps, 222
summary, 227
Resistance spot welding, 190, 198

S
Sliding friction
coefficient of friction, 128
relative velocity, 128
thermoelastic instability, 131
Solid spot conductance
defined, 2
Solid spot thermal conductance
correlations, 39
elastic deformation, 35
estimating by discretization, 44
fully plastic deformation, 34
numerical example, 42
Specific materials, 217, 234
correlations
stainless steel and aluminium, 234
zircaloy-2/uranium dioxide, 234
porous materials, 235
Stacks of laminations
correlation, 232, 233
Stretch blow moulding, 195
Summary, 175
other interstitial materials, 175
Surface profile
height, 27, 31, 47
slope, 27, 29

T
Thermal boundary resistance
acoustic mismatch model (AMM), 50

diffuse mismatch model (DMM), 50
Thermal conductivity
harmonic mean, 25
Thermal contact conductance
defined, 2
Thermal contact conductance of coated surfaces
effective hardness, 151
optimum thickness, 156
Thermal contact resistance
defined, 2
Thermal pastes, 170, 171, 174, 176
carbon black, 171, 173
metal particle, 171
Thermal spray coating, 196, 199
Transient measurements
conclusions, 90
Transient methods
laser flash method, 89

V
Variation of constriction resistance with time, 89
Variation of hardness, 49, 219
Variation of surface parameters, 35
Variation of thermal conductivity, 197

W
Waviness number, 38
Wire screens, 7, 139, 159, 247

9783319012759